PRAISE FOR ANDREW HINTON

From "Essential Math for AI:"

Pretty good outline of the math you'll need to get started if you need a refresher course.

— BOB MCGEE

Each topic is thoroughly examined to show how mathematical concepts apply to AI. The practical chapter summaries are invaluable for reinforcing learning and refreshing knowledge.

— ASHLEY GOODWIN

The author has done a remarkable job of compiling a guide that is not only informative but also incredibly practical. The application of each mathematical principle is carefully explained, demonstrating their relevance in developing robust AI systems.

— MAXWELL DB

BASIC MATH FOR AI

BASIC MATH FOR AI

A BEGINNER'S QUICKSTART GUIDE TO THE MATHEMATICAL FOUNDATIONS OF ARTIFICIAL INTELLIGENCE

AI FUNDAMENTALS
BOOK 5

ANDREW HINTON

Book Bound
STUDIOS

To all the curious minds eager to explore the world of Artificial Intelligence—may this book be your guide and companion on your journey to mastering the essential mathematics that power AI.

Mathematics is the language in which God has written the universe.

— GALILEO GALILEI

CONTENTS

READER BONUS

Thank you for purchasing this book! Get your copy of Andrew Hinton's limited edition book titled "The Power of AI."

It's a must-read for anyone interested in the future of technology and its impact on our world.

Visit bookboundstudios.wixsite.com/andrew-hinton or scan the QR code below.

SCAN ME

INTRODUCTION TO BASIC MATH FOR AI

Why Math is Essential for AI

Have you ever wondered why math is essential for artificial intelligence (AI)? Let's explore. At its core, AI involves programming computers to make decisions that usually require human intelligence. These decisions are based on complex algorithms and models that process vast amounts of data. Here's where math comes in—it's the backbone of these algorithms and models.

Think about it this way: without math, AI would be like trying to write a book without knowing the alphabet. You need math to quantify and interpret the world data that AI systems learn from. For instance, statistics help in understanding and predicting patterns from data. At the same time, calculus is used in optimizing these predictions. Linear algebra, on the other hand, is fundamental for data representation and manipulation in high-dimensional spaces, which is typical in AI applications.

Moreover, probability is pivotal in making decisions under uncertainty, a common scenario in AI systems. It helps assess the likelihood of various outcomes, which is crucial for tasks like speech recognition, language translation, and even autonomous driving.

In essence, math provides the tools to create models to learn from data, make predictions, and improve over time. It's not just about crunching numbers; it's about using those numbers to make intelligent decisions. So, if you're venturing into AI, a solid grasp of basic math is not just helpful—it's indispensable.

Overview of the Book

Welcome to "Basic Math for AI"! This book is designed to bridge the gap between fundamental mathematical concepts and their applications in artificial intelligence. Whether you're a student, a professional, or simply an AI enthusiast, this guide aims to equip you with the mathematical foundations necessary to understand and work effectively with AI technologies.

First, let's talk about the structure of the book. Each chapter is crafted to build on the knowledge from the previous one, starting with the most fundamental concepts and gradually moving to more complex topics.

This means that even if you're not a math wizard, you can follow along and grow your skills as you progress through the chapters.

To get the most out of this book, start at the beginning and work your way through each chapter in order. This approach will help you build a solid foundation and ensure you don't miss any key concepts crucial for understanding later material.

So, grab your calculator, open up your notebook, and start this exciting journey into the world of AI!

Introduction to Mathematical Thinking

Before we dive into the basics of math for AI, it's crucial to understand mathematical thinking. It extends beyond mere numbers and equations; it's a way of reasoning, problem-solving, and communicating.

Mathematical thinking in AI involves various mathematical disciplines such as algebra, calculus, statistics, and probability. Each area provides tools to model and solve real-world problems by transforming them into computable forms. This transformation is fundamental because, at its core, AI is about making sense of complex data and making predictions or decisions based on that data.

For instance, consider a simple AI application like a recommendation system on a streaming platform. The system uses algorithms to predict what you might like to watch next based on your viewing history. Behind this seemingly straightforward process is a complex mathematical framework involving statistics to analyze your past behavior, probability to predict future likes, and algebra to calculate similarities between different shows.

Moreover, mathematical thinking fosters a critical mindset for AI: the ability to abstract and generalize. When you abstract, you focus on the essential features of a problem, ignoring irrelevant details. This skill is vital in AI, where you often need to design algorithms that are not just solutions to a single problem but capable of handling various scenarios.

Conversely, generalization involves applying solutions from specific problems to a broader set of problems. This is seen in machine learning, a subset of AI, where an algorithm trained on a particular dataset can apply its learned insights to new, unseen data.

Embracing this mathematical mindset helps us understand AI and innovate within the field. As AI continues to evolve, mathematical models and theories also need to adapt and improve. This dynamic interplay between math and AI pushes the boundaries of what machines can do and continually reshapes our understanding of what mathematics can explain about the world.

So, as we move forward, remember that developing your mathematical thinking is not just about learning formulas and algorithms. It's about cultivating an analytical mindset that can see beyond the numbers, perceive patterns in the data, and innovate solutions as creative as they are computational. This journey into mathematical thinking is not just preparation for understanding AI—it's a way to think about the world.

1

FUNDAMENTALS OF ALGEBRA

Basic Algebraic Operations

Welcome to the world of algebra! This is where we see how numbers and letters can play together to solve problems. Think of algebra as a tool that helps us find unknowns, often represented by letters like x or y. It's like a treasure hunt where x marks the spot, and our job is to figure out what x is!

First, discuss the basic operations: addition, subtraction, multiplication, and division. These are the building blocks of algebra. We consider combining or taking away like terms when we add or subtract algebraic expressions. For example, if you have $3x + 2x$, you can combine these to get $5x$. It's like adding apples; if you have three apples and someone gives you two more, you now have five.

Multiplication in algebra looks trickier but follows the same principles as multiplying numbers. If you have *3x* and you multiply it by *2*, you get *6x*. Sometimes, you'll see expressions like:

$$(x + 2)(x + 3)$$

You'll use the distributive property to expand this into:

$$x^2 + 5x + 6$$

For reference, the distributive property states that *a(b + c)* is equal to *ab + ac*, which allows you to multiply each term inside the parentheses by the term outside.

Division, or splitting things into smaller parts, is also straightforward. If you have *6x* divided by *2*, you get *3x*. When dividing terms with the same base, you subtract the exponents, like in x^5 / x^2, which simplifies to x^3.

Now, let's remember solving equations, a central part of algebra. When you solve an equation, you're finding the value of the unknown that makes the equation true. For instance, if $x + 3 = 5$, you can solve for *x* by subtracting *3* from both sides of the equation, giving you $x = 2$.

Remember, the goal of algebra is to isolate the variable (the *x* or *y* or whatever letter is being used) on one side of the equation. This way, you can see what the unknown equals. You can tackle more complex problems and equations using these operations, setting a solid foundation for more advanced math needed in AI.

So, keep these tools handy as we dive deeper into algebra. They'll be invaluable as we explore how to model and solve problems, especially in artificial intelligence, where algebra helps create algorithms that can learn and make decisions. Let's get those algebraic gears turning!

Equations and Inequalities

Equations and inequalities are the bread and butter of algebra. They're fundamental, and understanding them is crucial, especially in fields like AI, where mathematical models dictate everything from machine learning algorithms to data analysis.

Let's start with equations. Think of an equation as a balance scale. Whatever you do to one side, you must do to the other to keep it balanced. This balance helps us find unknown values, often represented by variables like x or y. Remember the example in the previous section? In the equation $x + 3 = 5$, we can solve for x by subtracting 3 from both sides, giving us $x = 2$.

Now, inequalities are different. They tell us about the relative size of values, using symbols like $>$ (greater than), $<$ (less than), $\geq$ (greater than or equal to), and $\leq$ (less than or equal to). For example, $x + 3 > 5$ tells us that whatever value x represents, when you add 3 to it, the result is greater than 5. Solving inequalities involves similar steps to solving equations. Still, you must be careful about one thing: if you multiply or divide by a negative number, you must flip the inequality sign.

Equations and inequalities are powerful algebraic tools that help us describe and solve problems. In AI, they are used to set parameters, define constraints, and model real-world scenarios. Mastering them not only boosts your math skills but also deepens your understanding of how AI algorithms work, making you better equipped to handle the complexities of artificial intelligence.

Functions and Graphs

Understanding functions and their graphical representations is a cornerstone of algebra, especially when we consider Artificial Intelligence (AI). In its most basic form, a function is a relationship between a set of inputs and a set of outputs, where each input is related to exactly one output. This concept is a mathematical abstraction and a fundamental component of programming and AI algorithms.

Let's break it down with a simple example. Imagine you're programming a thermostat to control the temperature in your home. The function is the relationship between the time of day and the temperature setting. For instance, you might set the thermostat to 68 degrees Fahrenheit at night and 72 degrees during the day. Here, the input (time of day) determines the output (temperature setting), which is a practical function application.

Graphs, on the other hand, provide a visual representation of these relationships. They allow us to understand how changes in input values affect the output quickly. In our thermostat example, plotting a graph with time on the x-axis and temperature on the y-axis would show a step-like movement between 68 and 72 degrees, clearly illustrating the changes in temperature settings throughout the day.

My math teacher (Mr Wallis) taught me an example in high school algebra class. Mr Wallis used the example of baking bread, where the function related the amount of flour used to the bread loaf size. More flour equals larger bread. This visual and practical approach made a seemingly complex concept much more tangible and easier to understand.

Understanding functions and graphs in AI is crucial because algorithms often depend on these relationships to make decisions. For instance, in machine learning, functions can help predict outcomes based on input data. A well-plotted graph can reveal patterns in data that might not be obvious from raw numbers alone, aiding in the refinement of algorithms.

As we move forward, remember that the beauty of

functions and graphs lies in their ability to simplify complex relationships into understandable and actionable insights. Whether you're adjusting a thermostat or programming an AI to recognize speech patterns, the fundamental principles of functions and their graphical representations hold the key to unlocking a world of possibilities.

Polynomials

Polynomials are one of the building blocks of algebra, essential for anyone diving into AI and machine learning. Think of them as a way to express mathematical ideas succinctly, using a combination of variables and coefficients arranged in powers.

A polynomial can be as simple as $x + 2$ or as complex as:

$$4x^5 - 3x^3 + 2x^2 - x + 7$$

The highest power of the variable x (in this case, 5) determines the polynomial degree. This can tell us much about the function's behavior, especially when graphing it.

Why are polynomials important in AI? Well, they come into play in numerous algorithms, especially in areas like optimization, where you need to find the minimum or maximum of a function. They also appear in neural networks as polynomial activation functions, which help decide whether a neuron should fire.

Understanding how to manipulate polynomials is crucial. This includes operations like addition, subtraction, multiplication, and even division. For instance, adding $x^2 + 2x + 1$ and $3x^2 - x + 4$ results in $4x^2 + x + 5$. Each operation follows specific rules that maintain the structure of the polynomial, ensuring that the output is still a polynomial.

Another key aspect is factoring polynomials, breaking them down into simpler, irreducible pieces. This is particularly useful in solving polynomial equations, a common task in many AI applications. For example, factoring $x^2 - 5x + 6$ gives us $(x - 2)(x - 3)$, revealing that the solutions to $x^2 - 5x + 6 = 0$ are $x = 2$ and $x = 3$.

In summary, polynomials aren't just abstract mathematical concepts. They are practical tools that help model and solve real-world problems in artificial intelligence. By mastering polynomials, you're equipping yourself with the knowledge to tackle more complex algorithms and functions that you'll encounter in your AI journey.

Exponents and Logarithms

Let's dive into the world of exponents and logarithms, two fundamental concepts in algebra that play a crucial role in the mathematical underpinnings of artificial intelligence (AI). Understanding these concepts not only helps in solving complex equations but also in various algorithms and data processing techniques used in AI.

Starting with exponents is a way to express repeated multiplication of the same number. For example, 5^3 (read as "five raised to the power of three" or "five to the power of three") means multiplying 5 by itself three times:

$$5 * 5 * 5 = 125$$

Exponents are particularly useful in AI for operations involving powers of matrices and data scaling.

Now, logarithms are essentially the inverse operation of exponentiation. They answer: "To what power must the base be raised to produce a given number?" For instance, if you have $2^3 = 8$, the logarithm of *8* with base *2* is *3*, written as:

$$\log_2(8) = 3$$

In the context of AI, logarithms are invaluable for tasks like transforming nonlinear relationships into linear ones and simplifying data modeling and analysis.

Both exponents and logarithms have special rules that simplify calculations.

For exponents, these rules include:

- The product rule: $a^m * a^n = a^{(m+n)}$

- The quotient rule: $a^m / a^n = a^{(m-n)}$
- The power rule: $(a^m)^n = a^{(mn)}$

For logarithms, the corresponding rules are:

- The product rule: $log_b(MN) = log_b(M) + log_b(N)$
- The quotient rule: $log_b(M / N) = log_b(M) - log_b(N)$
- The power rule: $log_b(M^n) = nlog_b(M)$

These rules simplify manual calculations and enhance computational efficiency in AI applications, where handling large datasets and complex calculations is routine.

In summary, exponents and logarithms are not just abstract mathematical concepts but are tools that equip AI with the necessary computational power to perform tasks ranging from simple data processing to complex algorithm executions. As we continue exploring algebra's fundamentals, remember how these concepts interlink with real-world AI applications, making them exciting and immensely practical.

Chapter Summary

- Algebra introduces basic operations like addition, subtraction, multiplication, and division using variables represented by letters such as x or y.

- Combining like terms in algebra simplifies expressions, e.g., $3x + 2x$ becomes $5x$, similar to adding numbers.
- Multiplication in algebra can involve expanding expressions using the distributive property, e.g., $(x + 2)(x + 3)$ expands to $x^2 + 5x + 6$.
- Division in algebra involves simplifying expressions by dividing coefficients and subtracting exponents of like bases.
- Solving algebraic equations involves isolating the variable to find the value that makes the equation true, e.g., solving $x + 3 = 5$ to find $x = 2$.
- Inequalities use symbols to compare values and require careful operation handling, especially reversing the inequality sign when multiplying or dividing by a negative number.
- Understanding functions and their graphs is crucial in algebra, representing relationships between inputs and outputs.
- Polynomials, sequences of variables and coefficients, play a significant role in AI, especially in optimization and neural networks, and require operations like addition, subtraction, and factoring.

2

GEOMETRY ESSENTIALS

Basic Geometric Shapes and Properties

Let's dive into basic geometric shapes and their properties, a fundamental aspect of mathematics that plays a crucial role in artificial intelligence (AI).

First up, we have the simplest of shapes: the circle. Defined by its symmetry around a central point, every point on a circle is the same distance from the center. This distance is known as the radius. A key property of the circle is its circumference, the total distance around the circle, which can be calculated using the formula $C = 2\pi r$, where r is the radius.

Next, let's talk about triangles, the most basic polygons, and some of the most versatile. Triangles are classified into several types based on their side lengths and angles. Equilateral triangles have all sides and angles equal, making them highly symmetrical. Isosceles triangles have at least two equal sides, and scalene triangles have no equal sides. In terms of angles, a right triangle has one angle of 90 degrees, which is essential in many geometric calculations.

Squares and rectangles are the most recognizable quadrilaterals. A square has all sides equal and every angle at 90 degrees. A rectangle also has angles at 90 degrees, but only the opposite sides are equal. These shapes are everywhere, from the architecture we admire to the devices we use daily. They serve as a great example of how geometry applies to technology, particularly in graphical computations where their properties ensure stability and symmetry.

Those interested in more dynamic applications should consider the properties of shapes like the parallelogram or trapezoid. Parallelograms have opposite sides that are parallel and equal in length, making them useful in vector space studies and matrix algebra—both critical in advanced AI algorithms. Trapezoids, which have only one pair of parallel sides, often come into play in AI's graphical methods and integration techniques for optimizing functions.

Understanding these basic shapes and their properties equips us with the tools to tackle more complex geometric problems. Whether it's optimizing an algorithm's efficiency or improving a predictive model's accuracy, geometry principles are at the heart of many AI applications. So, as we move forward, remember that these aren't just shapes; they're the building blocks of the digital world, helping us to navigate and shape the future of technology.

Coordinate Geometry

Diving into the world of coordinate geometry, we find ourselves at the intersection of algebra and geometry, a key area in understanding how to navigate spaces. Coordinate geometry, or analytic geometry, allows us to describe geometric shapes numerically using coordinates and equations. This is particularly useful in AI for image recognition, computer graphics, and spatial analysis tasks.

At the heart of coordinate geometry is the Cartesian coordinate system. Introduced by René Descartes, this system uses two axes, horizontal (x-axis) and vertical (y-axis), intersecting at the origin. The location of any point in the plane can be described by an ordered pair of numbers (x, y), known as coordinates. The x-coordinate indicates the position along the horizontal axis. In contrast, the y-coordinate indicates the position along the vertical axis.

One of the fundamental concepts in coordinate geometry is the equation of a line, typically expressed in the form $y = mx + b$. Here, 'm' represents the slope of the line, which measures the steepness and direction, and 'b' represents the y-intercept, or the point where the line crosses the y-axis. Understanding how to manipulate and interpret this linear equation is crucial for developing algorithms that can, for example, predict trends from data or find patterns.

Moreover, coordinate geometry extends beyond straight lines to include shapes like circles, ellipses, and hyperbolas, each represented by specific equations. For instance, the equation of a circle with center at (h, k) and radius r is $(x - h)^2 + (y - k)^2 = r^2$. These equations become tools for AI in tasks such as object detection and motion tracking, where understanding the shape and position of different objects is essential.

In AI, the ability to convert geometric shapes into mathematical equations enables computers to process and analyze vast amounts of visual data efficiently. Whether for autonomous vehicles interpreting road signs or facial recognition systems identifying features, coordinate geometry provides the mathematical foundation for these complex tasks.

In summary, coordinate geometry is not just about plotting points and lines on a graph. It's about creating a bridge between abstract mathematical theories and real-world applications in AI, making it a fundamental topic for anyone venturing into this field. Understanding its principles allows for developing algorithms that can interpret and interact with the three-dimensional world around us, a critical skill in the era of automation and artificial intelligence.

Lines, Angles, and Their Relationships

Understanding the relationships between lines and angles is fundamental in geometry. AI often relies on these geometric principles to make sense of data, recognize patterns, and make decisions. Let's break down these concepts into digestible parts.

First, consider the humble line. In geometry, a line is straight and extends infinitely in both directions. It's characterized by its length, which is infinite, and has no curvature. When we talk about lines in the context of AI, think about how algorithms might use linear data to predict trends or behaviors.

Angles, on the other hand, are formed when two lines intersect. The space between these lines is measured in degrees, and this measurement is what we call an angle. Angles are pivotal in programming AI for object detection and navigation, where understanding the angle between objects can influence a machine's decision-making process.

Now, let's explore some specific relationships between lines and angles that are particularly useful:

1. **Parallel Lines:** These lines in a plane never meet; no matter how far they extend, they do not intersect. In AI, understanding parallel lines can help create parallel algorithms that

process data simultaneously, enhancing efficiency.

2. **Perpendicular Lines:** These lines intersect at a right angle (90 degrees). Recognizing perpendicular lines can be crucial in AI applications such as robotic movement, where angles of movement need to be precise.
3. **Transversal Lines:** A transversal is a line that passes through two lines in the same plane at two distinct points. This interaction creates several angles, some congruent (equal in measure). In AI, identifying these angles can help calibrate sensors that rely on angular measurements.
4. **Angle Pairs:** When a transversal intersects parallel lines, several angle pairs are formed, including corresponding angles, alternate interior angles, and alternate exterior angles. These relationships are essential in many AI applications, from visual recognition systems that need to align images correctly to spatial reasoning tasks where the AI must navigate or manipulate objects in three-dimensional space.

Understanding these concepts allows AI systems to interpret the world more accurately and interact with it more effectively. Whether a self-driving car calculates the safest path or a robotic arm precisely aligns components on an assembly line, the principles of lines and angles are deeply embedded in the algorithms that drive these technologies.

In summary, the study of lines and angles isn't just about drawing figures on a piece of paper; it's about laying the groundwork for intelligent systems that can think, learn, and interact with their environments in previously thought impossible ways.

Surface Area and Volume

Surface area is the total area that an object's surface occupies. It's like wrapping a gift and measuring the wrapping paper you need. Volume, on the other hand, deals with the amount of space an object occupies. Think of it as how much water you could pour into a container without spilling over.

Let's break it down with a simple example: consider a cube. If each side of the cube measures two units, the surface area would be six times the area of one face (since a cube has six faces), giving us a total surface area of 24 square units. The volume would be the length times the width times the height, which in this case is $2 * 2 * 2 = 8$ cubic units.

Now, why is this important in AI? Imagine you're programming a robot to pack boxes into crates. The robot must calculate the surface area to determine how much tape is required to secure the boxes and the volume to ensure the boxes fit perfectly in the crate without wasting space.

In a previous lifetime, I was heading up a project where we had to program a drone to map an archaeological site photographically. Calculating the surface area the drone needed to cover and its flight path volume was crucial for optimizing its battery life and the overall efficiency of the mapping process. This practical application of geometry helped preserve the site's integrity while capturing all necessary data.

So, whether optimizing logistics, planning efficient paths for robots, or creating simulations, understanding the geometry of surface area and volume can provide AI systems with the necessary information to make intelligent decisions. This foundational knowledge not only supports various functionalities in AI but also enhances the technology's effectiveness in real-world applications.

Geometric Transformations

Transformations play a pivotal role in geometry. Essentially, geometric transformations involve moving or changing geometric shapes in a specific manner. These transformations are categorized mainly into four types: translations, rotations, reflections, and dilations.

Starting with translations, imagine sliding a shape across the plane without rotating it or flipping it over.

This shift doesn't alter the shape or size of the figure; it merely repositions it. In mathematical terms, every point of the shape moves the same distance in the same direction.

On the other hand, rotations spin the shape around a fixed point, known as the center of rotation. The rotation angle determines how far the shape turns, and it's crucial to note that the shape does not change its size or form through this process. This transformation is akin to turning a key in a lock, where every part of the key moves around the central shaft.

Reflections are like looking into a mirror. Here, every point of a shape has an image on the opposite side of a line, known as the line of reflection. This transformation results in a mirror image of the original shape, pivotal in various symmetry operations in both natural and digital worlds.

Lastly, dilations involve resizing a shape by a specific scale factor. This transformation enlarges or reduces a shape but keeps its proportions and orientation intact. It's like zooming in or out on a camera, where the entire scene grows larger or smaller while maintaining the overall view.

Understanding these transformations is not just academic; they are heavily utilized in AI for image recognition, where an AI must recognize objects regardless of orientation, position, or size. By mastering geometric transformations, AI algorithms can be trained to have spatial awareness, indispensable in fields like autonomous driving, robotic surgery, and augmented reality.

In summary, geometric transformations are fundamental in manipulating and understanding shapes and spaces in pure mathematics and its applications in technology and AI. They allow us to alter the environment in a controlled manner, which is essential for creating intelligent systems that interact with the real world.

Chapter Summary

- Basic geometric shapes and properties are crucial in AI, especially in computer vision, robotics, and machine learning.
- Circles: Defined by a central point with equidistant perimeter points; key properties include radius and circumference.
- Triangles: Classified by side lengths and angles (equilateral, isosceles, scalene, right); fundamental in geometric calculations.
- Quadrilaterals: Symmetrical shapes like squares and rectangles are important in graphical computations.
- Advanced shapes: Parallelograms and trapezoids are relevant in vector spaces and integration techniques in AI.

- Coordinate geometry: Bridges algebra and geometry; essential for image recognition and spatial analysis using the Cartesian coordinate system.
- Lines and angles: Significant for object detection and navigation; includes concepts like parallel and perpendicular lines.
- Geometric transformations: Understanding translations, rotations, reflections, and dilations is crucial for algorithms to recognize objects despite changes in orientation, position, or size.

3

UNDERSTANDING CALCULUS

Introduction to Calculus

Welcome to the world of calculus! Often perceived as a challenging field of mathematics, calculus is a fascinating subject that offers profound insights into how our world operates. It is the mathematics of change and motion, and it plays a crucial role in various scientific disciplines, including artificial intelligence (AI).

At its core, calculus is divided into two branches: differential calculus and integral calculus. Differential calculus concerns the concept of a derivative, which can be considered a way to look at how a function changes at any given point. It's like zooming in on a curve to see how steep it is at any particular spot. This is incredibly useful in AI for optimizing algorithms—finding the best settings quickly and efficiently.

On the other hand, integral calculus deals with the concept of an integral, which is essentially about accumulation. If you've ever wondered about the total area under a curve, integral calculus helps you find that. This aspect is particularly useful in AI for processes like data smoothing and aggregation, where you must accumulate data over time to make predictions or adjustments.

Both concepts sound abstract now, but as we delve deeper into this chapter, we'll explore how they are applied in real-world scenarios, particularly in AI. We'll see how calculus helps in training algorithms to learn and make decisions, enhancing the intelligence aspect of artificial intelligence.

Limits and Continuity

Understanding limits and continuity in calculus is akin to learning how to read the basic alphabet before diving into complex literature. Let's simplify these concepts.

Imagine you're walking towards a doorway. As you get closer and closer, you're approaching what we call a "limit." In mathematical terms, a limit is the value that a function (think of it as a mathematical expression) approaches as the input (or x-value) approaches some value. Limits are essential because they help us handle situations where the function might be undefined, such as division by zero.

Now, let's talk about continuity. A function is continuous if, as you gently trace the curve with your finger, you don't have to lift your finger off the paper until you reach the end of the curve. More formally, a function is continuous at a point if the limit is equal to the function's value at that point. This means there are no breaks, jumps, or holes in the function at that point.

Why does this matter in AI? When algorithms make predictions based on continuous data, understanding where these data functions are continuous or where limits exist can help improve the accuracy of these predictions. For instance, in machine learning models like neural networks, calculus helps optimize the functions to reduce errors in predictions through a process called gradient descent.

In summary, grasping the concepts of limits and continuity equips you with foundational tools to delve deeper into calculus, paving the way for more advanced applications in artificial intelligence. Whether it's optimizing algorithms or understanding data trends, these concepts form the bedrock of many AI operations.

Differential Calculus

Differential calculus is a fascinating and essential branch of mathematics. It revolves around the concept of the derivative, which measures how a function changes as its input changes. This sounds a bit abstract at first, but it's incredibly practical. Think of it as finding the instantaneous rate of change or the slope of a curve at any point.

Let's break it down with a simple example. Imagine you're tracking the speed of a car as it accelerates. If you graph the car's speed over time, the slope of this graph at any point gives you the car's acceleration at that specific moment. In mathematical terms, if speed is a function of time, the derivative of this function at any point tells you the acceleration.

In AI, understanding how things change can help optimize performance. For instance, when training neural networks, differential calculus is used to find the minimum of a loss function. This process, known as gradient descent, involves calculating derivatives to determine the direction and rate at which the network's weights should be adjusted to minimize error.

The derivative is calculated using limits, a concept discussed in the previous section. The basic idea is to approximate the rate of change by looking at smaller and smaller intervals around a point. Mathematically, if you have a function *f(x)*, its derivative *f'(x)* at a point *x* is defined as:

$$f'(x) = \lim_{(h\to 0)} (f(x + h) - f(x)) / h$$

This formula might look daunting, but it's essentially

just a way to precisely define the slope of the tangent line to the function at point x.

But why is this important in AI? AI often involves making predictions, which can constantly be improved. By understanding how small changes in input (like adjusting parameters in an algorithm) can affect outcomes, we can fine-tune our AI models to perform better. This is where differential calculus plays a crucial role.

Moreover, differential calculus isn't just about finding the rate of change; it also helps in understanding the behavior of functions. For example, by examining the second derivative of a function, you can determine whether the function is curving up or down at a given point. This can indicate whether you've found a minimum or maximum value, which is crucial in optimization problems common in AI.

In summary, differential calculus provides powerful tools for navigating and optimizing complex systems in AI. By understanding derivatives, you can control and improve the learning processes of AI models, leading to more accurate predictions and efficient operations.

Applications of Derivatives

Understanding the applications of derivatives can be quite a game-changer in artificial intelligence. Derivatives, a fundamental concept in calculus, determine how a function changes at any given point. This is crucial in AI for optimizing algorithms, particularly those involving machine learning models.

One of the primary uses of derivatives in AI is in training neural networks through a process known as backpropagation. This method involves calculating the derivative of the loss function (a measure of how far the model's predictions are from the actual values) with respect to the network's weights. By understanding how the loss function changes with slight weight adjustments, developers can tweak these weights to minimize errors, improving the model's accuracy.

Another significant application is in computer vision, where derivatives help in edge detection in digital images. By applying derivatives, AI systems can identify sharp changes in intensity corresponding to the edges of objects in an image. This capability is fundamental in various applications, from autonomous vehicles needing to recognize obstacles to medical imaging systems that highlight important features in diagnostic images.

Moreover, derivatives are instrumental in reinforcement learning, a type of machine learning in which an algorithm learns to behave in an environment by performing certain actions and receiving rewards. The derivatives help optimize the reward function, ensuring that the model learns the most effective strategies over time.

In optimization problems, derivatives can find

minimum or maximum values. This is particularly useful in operations research and logistics. AI systems can help route planning, scheduling, and resource allocation by finding the most efficient solutions.

Understanding and applying derivatives allow AI systems to learn, adapt, and function precisely, mimicking human-like intelligence. This mathematical tool equips AI with the capability to perform complex tasks, making it an indispensable element in developing intelligent systems.

Integral Calculus

Integral calculus, often considered the counterpart to differential calculus, focuses on accumulating quantities and the spaces under curves. Where differential calculus cuts something into small pieces to find how it changes, integral calculus sums it together to determine how much there is.

One of the fundamental concepts in integral calculus is the definite integral. It's used to calculate the area under a curve between two points. It can be thought of as the accumulation of many tiny contributions. This concept isn't just a mathematical abstraction. For instance, in AI, integrating functions helps in areas like understanding the total error in a system over time or analyzing the probability of certain outcomes across a range of data.

The antiderivative, or indefinite integral, is another cornerstone of integral calculus. Unlike the definite integral, the antiderivative is more about finding a function whose derivative is the given function. This process is crucial in solving differential equations, which are fundamental in various AI models, especially those involving dynamic systems.

In AI, understanding how to integrate and differentiate functions allows for the optimization of algorithms and the effective training of models. For example, when training neural networks, integral calculus is used to adjust weights on inputs to minimize error rates and improve prediction accuracy.

Integral calculus might initially seem daunting, but its applications extend beyond the classroom. Whether optimizing functions in an AI algorithm or calculating risk probabilities in data sets, the integral binds together discrete data into a continuous and understandable whole. This integration—no pun intended—helps turn raw data into actionable insights, a process at the heart of artificial intelligence.

Applications of Integrals

Let's explore how integrals are applied in real-world scenarios relevant to AI.

Firstly, consider the area under a curve. This basic application of integrals is pivotal in AI for tasks like calculating probabilities and understanding distributions, which are essential for algorithms involving uncertainty and making predictions. For instance, in machine learning, integrals help determine the area under the ROC curve (AUC), a critical measure for evaluating the performance of classification models.

Another significant application is in computer vision, a field of AI that enables machines to interpret and process visual data. Integrals are used in image processing techniques to analyze images' features. For example, integral calculus is employed to compute the total brightness of an image or to perform transformations that help recognize patterns and shapes within the image.

In robotics, integrals play a crucial role in path planning and the navigation systems of autonomous robots. By integrating sensor data over time, robots can determine their position and orientation in space, enabling them to move accurately in their environment. This application is vital for developing self-driving cars, drones, and other autonomous systems.

Furthermore, in natural language processing (NLP), integrals are used in algorithms that analyze and interpret human language. They help in tasks such as speech recognition and machine translation by providing a way to aggregate continuous data points and effectively model linguistic patterns.

Lastly, integrals are instrumental in optimizing functions in AI algorithms. Many AI models, particularly deep learning ones, involve optimization techniques that require integration to minimize or maximize certain functions. This is crucial for training models to learn from data and make accurate predictions.

In summary, the application of integrals in AI spans various domains, from enhancing machine learning models and processing images to enabling autonomous navigation and optimizing functions. Understanding these applications provides insights into how integral calculus is utilized in AI. It underscores the importance of mathematical concepts in advancing technology.

Chapter Summary

- Calculus is divided into differential and integral calculus, which are essential for understanding change, motion, and accumulation in fields like AI.
- Differential calculus focuses on derivatives, which describe how functions change at specific points and help optimize AI algorithms.
- Integral calculus deals with integrals, which are used for accumulating quantities and calculating areas under curves, which is essential for data analysis in AI.
- Understanding limits and continuity in calculus is crucial for handling undefined

mathematical situations and ensuring smooth data functions in AI.

- Derivatives help AI by optimizing neural network training through processes like backpropagation and improving model accuracy.
- Applications of derivatives in AI include edge detection in computer vision and optimizing reward functions in reinforcement learning.
- Integral calculus applications in AI involve calculating probabilities, analyzing image features, and aiding in autonomous navigation.
- Both differential and integral calculus provide foundational tools for advancing AI technology through optimization and data interpretation.

4

PROBABILITY AND STATISTICS

Fundamentals of Probability

Let's explore probability, a fascinating and crucial area of mathematics, especially when it comes to artificial intelligence. Probability helps us quantify the uncertainty involved in predicting future events, making decisions under uncertainty, and modeling complex systems.

At its core, probability deals with the likelihood of different outcomes. To understand this concept, imagine flipping a fair coin. The probability of the coin landing on heads is 50%, just as the probability of it landing on tails is also 50%. This simple example introduces the idea of a probability space where the total probability of all possible outcomes (heads and tails in this case) adds up to 100% or 1.

Now, let's expand this concept. Consider rolling a six-sided die. The probability of rolling any specific number, say a 4, is 1 out of 6, or approximately 16.67%. Each outcome (1, 2, 3, 4, 5, or 6) has an equal chance of occurring because the die is fair. This scenario is a classic example of a uniform probability distribution, where every outcome has an equal likelihood of happening.

But not all probabilities are that straightforward. The outcomes are less likely in more complex situations, such as predicting the weather or the stock market, and the probability distribution can take different forms. For instance, the likelihood of rain might depend on factors like humidity, temperature, and wind conditions. Here, probabilities are calculated based on historical data and statistical models, which can get intricate.

In AI, understanding probability is essential for algorithms like Bayesian networks, which are used for decision-making and inferencing under uncertainty. These networks rely on probabilities to make educated guesses about unknown variables based on known variables.

Another critical concept in probability is independence. Two events are independent if the occurrence of one does not affect the occurrence of the other. For example, when flipping a coin twice, the outcome of the first flip does not influence the outcome of the second flip. Understanding independence is vital when analyzing complex systems where multiple factors interact.

As we explore AI applications further, we'll see how probability plays a role in machine learning, particularly in classification tasks and predictive modeling. Algorithms like logistic regression, for instance, use probability to estimate the likelihood of categorical outcomes, such as whether an email is spam or not.

In summary, probability offers a framework for dealing with uncertainty and making informed predictions and decisions. It's a foundational pillar in AI, enabling machines to learn from data and perform tasks that would otherwise require human intelligence.

Random Variables and Probability Distributions

Diving into random variables and probability distributions, we're exploring the backbone of how uncertainty and randomness are quantified in mathematical terms.

A random variable is a variable whose possible values are numerical outcomes of a random phenomenon. There are two types of random variables: discrete and continuous. Discrete random variables have specific, separate values, like the number of heads in a series of coin flips. Continuous random variables, on the other hand, can take any value within a range, such as the time it takes for a computer to process a certain algorithm, which could be any non-negative real number.

Each random variable has a probability distribution

describing how probabilities are assigned to each possible value. For discrete variables, this is often represented by a probability mass function (PMF). For continuous variables, we use a probability density function (PDF). These functions give us a framework to calculate the likelihood of various outcomes. They are fundamental in fields ranging from economics to engineering.

For instance, consider the previous example of rolling of a fair six-sided die. The probability of rolling any specific number is uniformly distributed since each outcome from 1 to 6 has an equal chance of occurring. This scenario can be described using a discrete uniform distribution. On the flip side, if we were measuring the amount of time until a light bulb burns out, this might follow a continuous distribution like the exponential distribution, where the probability of the bulb lasting a specific amount of time changes continuously.

Understanding these distributions allows AI systems to predict future events based on patterns observed in data. For example, in machine learning, probability distributions are used to handle data uncertainties and make inferences about the underlying processes generating the data. This could predict anything from the likelihood of a user clicking on an ad based on historical click data to an AI diagnosing diseases from medical images by learning from distributions of known outcomes.

In summary, random variables and their probability

distributions are not just abstract mathematical concepts but practical tools used in AI to deal with the real world's randomness and uncertainty. They help AI systems learn from the past and make informed predictions, indispensable in developing intelligent systems that perform reliably across various scenarios.

Statistical Measures

Understanding statistical measures is essential in artificial intelligence. These measures give us insights into data that can influence how an AI system is trained, performs, and can be improved. Let's examine some key statistical measures and see why they matter.

First up, we have the mean, often referred to as the average. It's calculated by adding all the values in a dataset and dividing by the number of values. The mean gives us a central value, which is handy when you want to get a general idea of the data's behavior without getting into the details.

Next, there's the median. This is a dataset's middle value when arranged in ascending order. If the number of observations is even, the median is the average of the two middle numbers. Unlike the mean, the median is unaffected by extraordinarily high or low values, making it particularly valuable in understanding the central tendency of skewed data.

Then, we have the mode, which is the value that

appears most frequently in a dataset. In some cases, data can have more than one mode or even no mode at all! The mode is handy in categorical data analysis, where we want to identify the most common category or value.

Let's remember variance and standard deviation. Variance measures how spread out the values in a dataset are. It's calculated by taking the average of the squared differences from the mean. On the other hand, the standard deviation is the square root of the variance and provides a measure of the spread of data points around the mean. In AI, understanding the variance and standard deviation can help in fine-tuning algorithms, especially in complex models where data dispersion influences performance.

Lastly, we touch on the concept of range. Range is the difference between the highest and lowest values in a dataset. It gives a quick sense of the spread of values but only tells you a little about the distribution between those extremes.

Each of these statistical measures has its place in AI. They help summarize data, which is crucial when deciding how to train models or adjust parameters. AI developers can better predict model behaviors and ensure more reliable outputs by understanding the central tendencies and variabilities.

Hypothesis Testing

Hypothesis testing is a fundamental statistical concept that allows us to decide about a population based on sample data. It's particularly crucial in AI, where making predictions and data-based decisions is a daily routine. Let's break down this concept into more digestible parts.

Imagine you're trying to determine whether a new algorithm improves the accuracy of a machine-learning model. You start with a hypothesis, a statement that you want to test. We call the initial hypothesis the null hypothesis in statistics, often denoted as H_0. This hypothesis represents a default position in which there is no effect or difference. In our example, the null hypothesis would be that the new algorithm does not improve accuracy.

Opposite the null hypothesis is the alternative hypothesis, denoted as H_1. It suggests that there is an effect or there is a difference. The alternative hypothesis for our algorithm would be that it improves the model's accuracy.

To test these hypotheses, you would collect sample data. For instance, you might implement the new algorithm in several projects and measure the model's accuracy. Then, using statistical tests, you can analyze the data to determine whether the results are significant enough to reject the null hypothesis and accept the alternative hypothesis.

One common method of hypothesis testing is the t-test, which assesses whether the means of two groups are statistically different. Another method is the chi-square test, used for categorical data to see if distributions of categorical variables differ.

The outcome of a hypothesis test is determined by the p-value, which helps you determine the significance of your results. The p-value indicates the probability of obtaining test results at least as extreme as the results observed under the assumption that the null hypothesis is correct. A small p-value (typically ≤ 0.05) indicates strong evidence against the null hypothesis, so you reject the null hypothesis.

A critical aspect of hypothesis testing in AI is understanding that it's not just about accepting or rejecting a hypothesis but about understanding the strength of the evidence. This is crucial when tuning algorithms or choosing between approaches in machine learning projects.

Regression Analysis

Regression analysis is a powerful statistical tool used to examine the relationship between two or more variables. Imagine you're trying to predict the price of a house based on its size, location, and age. Regression analysis helps you understand how these factors affect your prediction.

At its core, regression aims to fit a model to the data points. It estimates the model's coefficients to minimize the difference between the observed values and the values predicted by the model, which is known as fitting a regression line. The most common type of regression analysis is linear regression, where the model predicts a linear relationship between the dependent and independent variables.

For instance, if you're analyzing the impact of study hours on exam scores, linear regression could help you predict scores based on the number of hours spent studying. The equation for a simple linear regression model is:

$$Y = a + bX$$

Where *Y* is the dependent variable (exam scores), *X* is the independent variable (study hours), '*a*' is the intercept, and '*b*' is the slope of the line. The slope '*b*' tells you how much the exam scores are expected to increase for each additional hour of studying.

In the context of AI, regression analysis is indispensable for tasks like forecasting (think stock prices or weather), optimizing processes, and even training algorithms to understand complex patterns. For example, in machine learning, regression models can help predict outcomes based on historical data, which is crucial for features like recommendation systems or automated decision-making processes.

Moreover, regression analysis isn't limited to linear relationships. There are multiple forms, such as multiple regression, where several independent variables are used, and non-linear regression, which can model more complex relationships. Each type of regression has its specific use case depending on the nature of the variables and the relationship between them.

Understanding the basics of regression analysis enhances your data analysis toolkit and equips you with the insights to apply mathematical models effectively in real-world AI applications. Whether you're optimizing an AI's decision-making process or predicting future trends, regression analysis provides a foundation for making informed, data-driven decisions.

Bayesian Statistics

Bayesian statistics is a fascinating and influential branch that provides a framework for updating beliefs in light of new evidence. This approach is particularly useful in artificial intelligence, where making decisions based on incomplete or evolving data is a common challenge.

At the heart of Bayesian statistics is Bayes' Theorem, which mathematically describes how to update the probability of a hypothesis as more evidence or information becomes available. This theorem uses the prior probability of a hypothesis before seeing the data, the likelihood of the observed data given the hypothesis, and the marginal likelihood of the observed data under all hypotheses being considered.

Let's break it down with a simple example. Imagine you're developing an AI system to diagnose diseases based on symptoms. Initially, you have a belief (prior probability) about the prevalence of a disease. As you gather symptoms (data), Bayes' Theorem helps you update your belief to reflect this new information, resulting in a posterior probability. This posterior probability is a more informed belief about the likelihood of the disease.

One critical advantage of Bayesian statistics is its flexibility. You can continuously update the probabilities as new data comes in, making it ideal for AI applications where real-time decision-making is crucial. This contrasts traditional statistics, which use fixed data sets to make inferences.

Moreover, Bayesian methods handle uncertainty very explicitly. By working with probabilities, these methods allow AI systems to express confidence in their predictions, which can be crucial for applications like autonomous driving or medical diagnosis.

However, Bayesian statistics can be computationally intensive, especially with large data sets or complex models. This is because it requires integrating many parameters to calculate the marginal likelihood.

Fortunately, modern computational techniques and the increasing power of computers have made Bayesian methods more accessible and practical for a wide range of applications in AI.

In summary, Bayesian statistics offers a robust framework for dealing with uncertainty and incorporating new information, making it a valuable tool in developing intelligent systems. As AI continues to evolve, Bayesian statistics principles will play a crucial role in enabling machines to learn from data, adapt to new situations, and make informed decisions.

Chapter Summary

- Probability is essential in AI for quantifying uncertainty in predictions and decision-making, using examples like coin flips and dice rolls to illustrate basic concepts.
- Probability distributions, which can be uniform or vary based on factors like weather conditions, are crucial for modeling complex scenarios in AI.
- Independence in probability, where the outcome of one event does not affect another, is key for analyzing systems with multiple interacting factors.
- AI uses probability in machine learning for tasks like classification and predictive

modeling, with algorithms estimating the likelihood of outcomes.

- Random variables represent outcomes of random phenomena and are categorized as discrete or continuous, each with specific probability distributions.
- Statistical measures such as mean, median, mode, variance, and range help summarize data, which is crucial for training and improving AI models.
- Hypothesis testing in AI involves making decisions based on sample data, using t-tests and chi-square tests to compare hypotheses.
- Regression analysis predicts relationships between variables, crucial for AI tasks like forecasting and optimizing processes with linear and non-linear models.

5

DISCRETE MATHEMATICS

Set Theory

Set theory is a fundamental part of mathematics, particularly in discrete mathematics. It plays a crucial role in artificial intelligence (AI) by providing a framework for organizing and structuring data, which is essential for algorithms and programming in AI.

At its core, the set theory deals with the concept of a 'set,' which is simply a collection of distinct objects considered as a whole. These objects can be anything: numbers, symbols, points in space, etc. The beauty of set theory lies in its ability to deal with finite and infinite collections, making it incredibly versatile and powerful.

One of the first concepts we encounter in set theory is the notion of membership. If an object *'a'* is a member of a set *'A,'* we write:

$$a \in A$$

If *'a'* is not a member of *'A,'* we write:

$$a \notin A$$

This basic concept leads us to more complex ideas, such as subsets, unions, intersections, and complements. For instance, a set *'B'* is a subset of *'A'* (written as $B \subseteq A$) if every element of *'B'* is also an element of *'A.'*

Understanding how sets interact through operations like union (combining elements from two sets), intersection (elements common to both sets), and complement (elements in one set but not in another) is crucial. These operations help build and refine selections and conditions in programming, which are daily tasks in AI development.

Moreover, set theory introduces the concept of cardinality, which measures the number of elements in a set. This concept is particularly important in AI when dealing with large datasets, as it helps understand the complexity and limitations of data processing tasks.

In practical AI applications, set theory is used to design and optimize algorithms for searching, sorting, and managing databases. It also underpins the logic used in many AI models, helping to define relationships and constraints within data.

In summary, set theory is about more than just dealing

with collections of objects. It's about applying these concepts to solve real-world problems in AI, making it a critical area of study for anyone looking to excel in this exciting and evolving field. Whether you're programming a simple AI or developing complex algorithms, the principles of set theory will provide the foundation you need to think clearly and work effectively.

Logic and Boolean Algebra

Understanding the principles of logic and boolean algebra is not just useful; it's essential in the realm of artificial intelligence. These mathematical frameworks provide the backbone for structuring rational thought processes and decision-making in AI systems.

Let's start with logic. It's all about reasoning, and it comes in two flavors: deductive and inductive. Deductive reasoning involves deriving specific truths from general statements, while inductive reasoning involves making generalizations based on specific instances. In AI, these reasoning methods help machines make decisions and learn from data.

Boolean algebra is a bit like the binary language of computers. It revolves around variables that have two possible values: true or false. This might sound simplistic, but it's incredibly powerful. Boolean algebra uses operators like *AND*, *OR*, *NOT*, *XOR*, etc., to combine these true or false values to solve complex problems. For instance, a self-driving car uses boolean logic to decide whether to stop or go at a traffic light.

The beauty of boolean algebra in AI lies in its ability to simplify decision-making processes. By breaking down decisions into simpler, binary choices, AI systems can process vast amounts of information more efficiently and make decisions quickly. This is crucial in applications with high speed and accuracy, such as real-time systems.

Moreover, boolean functions are fundamental in designing electronic circuits and systems and integral to hardware running AI algorithms. Understanding how these functions work and how to manipulate them can lead to more efficient designs and, ultimately, more intelligent AI.

In summary, logic and boolean algebra are not just academic subjects but practical AI tools. They help machines reason, make decisions, and interact with the world in a way that mimics human thought processes.

Graph Theory

Graph theory is a fascinating and essential area of discrete mathematics. It deals with graphs and mathematical structures that model pairwise relations between objects. A graph comprises vertices (or nodes) and edges (or links) that connect pairs of vertices. In AI, graphs are used in various applications, such as neural networks and pathfinding algorithms, and for modeling relationships in complex systems.

One of the basic concepts in graph theory is the distinction between directed and undirected graphs. In directed graphs, the edges have a direction, indicating a one-way relationship. In contrast, the edges have no direction in undirected graphs, indicating a mutual relationship. This distinction is crucial in AI for designing algorithms that model different types of data relationships.

Another essential concept is graph connectivity, which refers to the paths linking pairs of nodes. Ensuring a network is adequately connected in AI can mean the difference between a robust, efficient system and one that fails to perform. Algorithms such as Depth-First Search (DFS) and Breadth-First Search (BFS) are fundamental in exploring these paths. They are widely used in AI for tasks like searching for data in a database or finding the shortest path in routing and navigation systems.

Graph theory also introduces the concept of cycles and acyclic graphs, which are graphs without cycles. Acyclic graphs are particularly important in AI for constructing hierarchical models, such as those used in machine learning for decision-making processes.

Moreover, the efficiency of AI algorithms often depends on graph coloring techniques, which involve assigning colors to the vertices of a graph under certain constraints. Graph coloring is used in scheduling algorithms and is pivotal in AI applications involving task allocation and resource management.

Understanding graph theory enhances the design and analysis of AI algorithms and contributes to more innovative solutions in AI applications. Whether it's optimizing routes, processing images, or predicting user behavior, graph theory provides a framework that supports complex data structures and algorithms essential for AI advancements.

Combinatorics

Combinatorics is another fundamental part of discrete mathematics. It deals with the study of finite or countable discrete structures. It involves counting, listing, and determining the existence of specific properties of arrangements or selections of objects. This field is not just about solving abstract mathematical problems; it's about applying these concepts to real-world scenarios, which can be incredibly useful in AI for optimizing algorithms, managing data structures, and even decision-making processes.

Let's break it down with a simple example: Suppose you're organizing a dinner party. You have ten friends but only room to invite five. Combinatorics helps you determine how many groups of five you can invite. This kind of problem, a combination, is just the tip of the iceberg. In AI, similar problems arise when selecting optimal subsets of data from larger datasets, a common task in machine learning models.

Permutations are another key concept in combinatorics. While combinations consider selecting items where order does not matter, permutations are all about ordering. Think about passwords: How many different six-character passwords can you create using letters and numbers? Each arrangement or order of characters counts as a unique permutation, and understanding this helps in areas of AI like cryptography and security.

In summary, combinatorics equips AI professionals with the tools to tackle scalability, efficiency, and complexity issues. By understanding how to count and arrange and applying these principles to data, AI can be more powerful and efficient.

Number Theory

Number theory often considered the purest of mathematics, is about the properties and relationships of numbers, especially integers. It's a fascinating field that combines the simplicity of basic arithmetic with the complexity of deeper analysis.

Let's start with the basics: prime numbers. These are the building blocks of number theory. A prime number is greater than one that cannot be formed by multiplying two smaller natural numbers. Primes are crucial in various encryption algorithms that keep our digital world secure.

Next, we have the greatest common divisor (GCD), the largest number that divides two integers without leaving a remainder. The Euclidean algorithm, an ancient method for computing the GCD, is efficient and a beautiful demonstration of how iterative processes can solve seemingly complex problems.

Modular arithmetic is a system of arithmetic for integers, where numbers "wrap around" upon reaching a specific value—the modulus. If you've ever used a 12-hour clock, then you've used modular arithmetic! This concept is vital in computer science, especially cryptography and coding theory.

Lastly, let's discuss Diophantine equations, named after the ancient mathematician Diophantus. These are polynomial equations for which we seek integer solutions. Although they may sound esoteric, they are central to many branches of mathematics and have applications in fields as diverse as number theory and algebraic geometry.

Each of these topics deepens our understanding of mathematics and enhances our ability to develop algorithms that can perform tasks ranging from simple sorting to complex data encryption. As AI continues to evolve, the role of number theory also expands, proving that the oldest branches of mathematics are sometimes still the most relevant.

Chapter Summary

- Set theory is crucial in discrete mathematics and AI, providing a framework to organize data essential for algorithms and programming.
- It involves 'sets,' collections of distinct objects, and operations like unions, intersections, and complements to manage data relationships.
- Cardinality, or the number of elements in a set, is significant in AI for understanding data complexity and processing limitations.
- Logic and Boolean algebra are foundational in AI, structuring rational thought processes and decision-making through operators like AND, OR, and NOT.
- Graph theory deals with vertices and edges in graphs. It is used in AI for modeling relationships and optimizing algorithms like neural networks and pathfinding.
- Combinatorics involves counting and arranging discrete structures, which is crucial for optimizing AI algorithms and managing data structures.
- Discrete probability helps model scenarios with distinct outcomes in AI, essential for

making decisions based on uncertain information.

- Number theory focuses on the properties of numbers. It is fundamental in AI to develop efficient algorithms, especially in encryption and security.

6

LINEAR ALGEBRA

Vectors and Spaces

In artificial intelligence, understanding vectors and spaces is like getting to know the building blocks of a language before you start forming sentences. Vectors are not just arrows pointing in space but are fundamental in defining directions and states in AI models. Think of them as a way to store and manipulate data. Each vector in an AI model could represent anything from an image's pixel values to consumer behavior features.

Vectors are usually listed in columns or rows and consist of elements from a field, typically real numbers. For instance, a three-dimensional vector might look like this:

$$[1, 3, 5]$$

Each element represents a coordinate in some space, dictating the direction and magnitude of the vector.

Now, when we talk about spaces in linear algebra, we're referring to vector spaces. A vector space can be considered the playground where all vectors of a certain type can roam freely. It's a mathematical structure formed by vectors that can be scaled and added to form new vectors.

The concept of vector spaces extends into subspaces, which are like smaller playgrounds within the larger one, where certain rules of vector operations still hold. For example, a plane through the origin forms a subspace in a three-dimensional space.

Why does this matter in AI? Well, operations on vectors and the manipulation of spaces form the backbone of machine learning algorithms. They allow us to perform calculations that can adjust predictions, optimize decisions, and, ultimately, make sense of vast amounts of data.

Understanding how to manipulate these vectors and spaces effectively allows AI systems to learn from data, recognize patterns, and make decisions—essentially, to "think" in a structured and logical way. This is crucial in developing algorithms that can perform tasks ranging from recognizing faces in photos to predicting stock market trends.

In summary, vectors, and spaces are not just abstract mathematical concepts but are practical tools in the toolkit of any AI developer. They help translate complex real-world data into a language that computers can understand and act upon.

Matrix Algebra

A matrix is a rectangular array of numbers arranged in rows and columns. Think of it as a spreadsheet, which is used to organize data. Still, in mathematics and AI, these matrices are more than just data storage—they are tools for performing complex calculations that can represent transformations and operations across multidimensional spaces.

One of the first operations we encounter with matrices is addition. Matrix addition is straightforward: you add corresponding elements from each matrix. For instance, if you have two matrices, A and B, of the same dimensions, their sum, C, is defined such that each element $C_{ij} = A_{ij} + B_{ij}$. It's like adding expenses in different categories from two months; you line them up and add the corresponding figures.

Matrix multiplication, however, is a bit trickier and far more powerful. It involves taking the rows of the first matrix and the columns of the second matrix, multiplying corresponding elements, and then summing them up to produce a new matrix. This operation is crucial because it allows for data transformation in ways that addition alone cannot achieve. For example, matrices rotate, resize, and transform images in graphics and image processing.

In AI, these concepts are extended to higher dimensions. For instance, when training a neural network, the data transformations and adjustments are handled through matrix operations. This allows the network to learn from vast amounts of data and make decisions or predictions. Understanding how these operations work gives us a glimpse into the inner workings of AI algorithms.

Moreover, matrices are not just about numbers. They can represent probabilities, state transitions, or any multidimensional data points. In AI, understanding the structure and manipulation of matrices is crucial for algorithms like Markov chains, which predict future states in a system.

In summary, matrix algebra is not just a topic of abstract mathematical theory but a practical tool that drives much of the technology in AI. From image processing to machine learning, the ability to manipulate matrices efficiently can lead to significant advancements and insights.

Determinants and Inverses

Understanding determinants and inverses is crucial in linear algebra, especially when solving systems of linear equations, which is a common task in artificial intelligence for optimizing algorithms and processing data.

Let's start with determinants. A determinant is a scalar value that can be computed from the elements of a square matrix and encodes certain matrix properties. One fundamental property is that it determines whether a matrix is invertible or not—a non-zero determinant indicates that the matrix has an inverse. In contrast, a zero determinant means it does not. This is vital because the invertibility of matrices is essential in many AI applications, including finding solutions to linear systems and performing transformations in graphics and deep learning.

Calculating the determinant of a matrix might seem daunting at first, but it's pretty straightforward with practice. For a 2x2 matrix, the determinant is simply $ad - bc$, where a, b, c, and d are the elements of the matrix arranged as follows:

[a, b]
[c, d]

For larger matrices, the process involves breaking down the matrix into smaller matrices, a method known as expansion by minors. While the calculations get more complex with the increase in matrix size, the fundamental concept remains the same.

Moving on to inverses, the inverse of a matrix A is

another matrix, denoted as $A^{(-1)}$, and it holds a unique property: when multiplied by *A*, it results in the identity matrix. The identity matrix is the equivalent of 1 in matrix algebra, meaning it does not change other matrices when it's multiplied with them.

To find the inverse of a matrix, you apply the formula:

$$A^{-1} = (1 / \det(A)) * \operatorname{adj}(A)$$

Here, *adj(A)* refers to the adjugate (or adjoint) of matrix *A*, which is obtained by taking the transpose of the cofactor matrix of *A*. Remember, this formula only works if the determinant of *A* is non-zero.

Understanding how to compute determinants and inverses is not just academic; it's a practical skill that helps in various AI tasks. For instance, in machine learning, inverses are used in algorithms like linear regression to calculate the weights that minimize prediction error. Similarly, in deep learning, determinants can help understand the behavior of transformations in neural network architectures.

By mastering these concepts, you're not just learning abstract mathematics; you're equipping yourself with the tools to tackle real-world problems in AI, making your journey into this exciting field informed and competent.

Eigenvalues and Eigenvectors

Diving into linear algebra, eigenvalues, and eigenvectors stand out as fundamental concepts that are pivotal in mathematics and essential in various applications within artificial intelligence. Let's break these concepts down into simpler terms and explore their significance.

Imagine a transformation represented by a matrix acting on a vector. In many cases, this transformation changes the direction of the vector. However, certain special vectors only get scaled (stretched or shrunk) by this transformation, not changing their direction. These particular vectors are called eigenvectors of the matrix. The factor by which they are scaled is known as the eigenvalue corresponding to that eigenvector.

To find a matrix's eigenvalues, we look for scalars, λ, such that when we subtract λ times the identity matrix from our original matrix and multiply by a vector, the result is the zero vector. This condition leads us to a characteristic equation, a polynomial equation for λ. The solutions to this equation, the roots, are the matrix's eigenvalues.

Once the eigenvalues are known, eigenvectors can be determined by substituting each eigenvalue back into the equation formed by subtracting the eigenvalue times the identity matrix from the original matrix and solving for the vectors that satisfy this equation. This process, though algebraically intensive, reveals the eigenvectors.

Why are these concepts so crucial in AI? In artificial intelligence, especially in areas like machine learning and data analysis, eigenvalues and eigenvectors are used to understand and compute principal components for dimensionality reduction in datasets. This technique, known as principal component analysis (PCA), helps reduce the complexity of data, improve the efficiency of algorithms, and enable the extraction of important features from large datasets.

Moreover, in neural networks, the eigenvalues of the matrices involved can often be used to analyze the stability and dynamics of learning algorithms. Eigenvalues can indicate whether specific learning processes converge correctly and how they can be optimized.

In summary, eigenvalues and eigenvectors are not just abstract mathematical concepts but tools that provide deep insights into data and algorithms in the realm of AI. They help simplify complex problems, make data more manageable, and ensure that AI systems operate efficiently and effectively.

Linear Transformations

In linear algebra, linear transformations are fundamental concepts that connect abstract theory with practical application. A linear transformation is a function between two vector spaces that preserves vector addition and scalar multiplication. This means that for any vectors u and v in a vector space, and any scalar c, the transformation T satisfies the following conditions:

1. $T(u + v) = T(u) + T(v)$
2. $T(cu) = cT(u)$

These properties ensure that the transformation is "linear," meaning it does not bend or curve the space in which the vectors reside. Instead, it may stretch, compress, or rotate the space.

To visualize this, imagine you have a digital image—a grid of pixels, each defined by a vector representing color and intensity. Applying a linear transformation to this image could rotate, resize, or shift it. Still, the grid structure remains consistent, and the relationships between the pixels are maintained.

In AI, linear transformations are extensively used in the training of neural networks. Each layer of a neural network typically performs a linear transformation on its input before applying a nonlinear activation function. The parameters of these transformations are adjusted during training to minimize the difference between the network's actual output and the desired output.

Moreover, linear transformations are represented by matrices. This representation is not just a convenient mathematical abstraction but is practically useful. Operations like rotations, reflections, and shearing in graphics programming are implemented using matrices. Each column of a transformation matrix is the image of a unit vector under the transformation, providing a clear geometric interpretation of the matrix.

Understanding how these transformations work and how to manipulate them is crucial for developing more efficient and effective AI algorithms. The ability to decompose and reconstruct transformations can lead to a more intuitive understanding of complex multi-layer networks, where each layer's transformation can be tweaked individually to optimize performance.

In summary, linear transformations are not just dry, abstract elements of mathematical theory. They are active, dynamic processes that shape data in the form of images, signals, and even the high-dimensional datasets typical in machine learning. They help us mold the digital representations of the world to fit the models we are building in AI, making them indispensable tools in the AI toolkit.

Applications to Machine Learning

Linear algebra is a powerhouse in machine learning, providing the mathematical framework that underpins many of the algorithms that drive artificial intelligence today. Let's explore how this branch of mathematics plays a pivotal role in developing AI technologies.

First off, consider the concept of vectors and

matrices, fundamental elements of linear algebra. In machine learning, vectors can represent anything from pixels in an image to words in a document. At the same time, matrices are often used to store data sets or weights in neural networks. This makes linear algebra essential for data preprocessing, transformations, and critical optimization in training models.

Training neural networks is one of the most common applications of linear algebra in machine learning. Here, matrix multiplication becomes a critical operation, allowing for the efficient calculation of outputs from multiple layers of neurons. The weights of these neurons, which determine how input data is transformed as it passes through the network, are adjusted during training to minimize the difference between the predicted output and the actual output, a process heavily reliant on linear algebraic operations.

Moreover, eigenvalues and eigenvectors, another topic we've touched upon, are crucial in methods like principal component analysis (PCA). PCA is used extensively for dimensionality reduction—simplifying data with many variables into principal components that retain the most important information. This is particularly useful in handling high-dimensional data sets, improving the efficiency and performance of machine learning models.

Linear transformations involve applying a function to all the points in a vector space to produce a new space.

These transformations are fundamental in understanding and constructing neural networks. They help learn complex patterns from data, which is essential for tasks such as image recognition, speech recognition, and many other AI applications.

In summary, linear algebra is not just a tool for mathematical manipulations. It's the backbone of machine learning, enabling systems to learn from data, make predictions, and improve autonomously. As we continue to push the boundaries of what AI can achieve, linear algebra remains foundational and indispensable. Whether you're tweaking a model to predict better market trends or developing algorithms that can diagnose diseases from medical images, linear algebra is your mathematical companion on this exciting journey into the future of AI.

Chapter Summary

- Vectors in AI represent data like image pixel values or consumer behavior features. They are fundamental in defining directions and states in AI models.
- Vector spaces in linear algebra are structures where vectors can be added and scaled, forming the basis for operations in machine learning algorithms.
- Matrix algebra is crucial in AI for organizing

data and performing complex calculations, such as transformations across multidimensional spaces.

- Determinants and inverses of matrices are essential in AI for solving linear systems and performing transformations, with non-zero determinants indicating invertibility.
- Eigenvalues and eigenvectors are used in AI for dimensionality reduction and understanding data transformations, which are crucial for efficient algorithm performance.
- Linear transformations in AI maintain vector operations and are represented by matrices. They play a pivotal role in neural network training and data manipulation.
- Linear algebra applications in machine learning include data preprocessing, neural network training, and dimensionality reduction using techniques like PCA.
- Linear algebra is foundational in machine learning. It enables systems to learn from data, make predictions, and improve autonomously, which is essential for advanced AI applications.

7

NUMERICAL METHODS

Numerical Integration and Differentiation

Diving into the world of numerical methods, mainly numerical integration and differentiation, is like unlocking a new level in a game where the challenges get more exciting and the tools you use become more sophisticated. These techniques are essential for solving real-world problems in artificial intelligence, where analytical solutions are impossible or impractical.

Let's start with numerical integration. Imagine you're trying to figure out the total area under a curve, representing the growth of a company's revenue over time. The curve is complex, and there's no simple formula for the area. Numerical integration allows you to approximate this area by breaking the curve into small, manageable pieces (like rectangles or trapezoids), calculating the area of these pieces, and then summing them up. Techniques such as the Trapezoidal rule, Simpson's rule, and Monte Carlo methods come into play here. Each has its perks and pitfalls, but all aim to give you a workable approximation that improves with finer subdivisions of the curve.

Now, switch gears to numerical differentiation. This is about finding the rate at which something changes. In AI, you need to know how quickly an algorithm is learning or how sensitive a prediction model is to changes in input. Numerical differentiation helps you approximate the derivative of a function at a given point when you can't derive it analytically. By calculating the slope of secant lines through points on the function that are close together, you can get a reasonable estimate of the derivative. Methods like forward difference, backward difference, and central difference are commonly used, each with its balance of accuracy and computational intensity.

Both these numerical methods are not just academic exercises; they are practical tools used in machine learning for optimizing algorithms, in computer vision for processing images, and in natural language processing for understanding and generating text. The beauty of these methods lies in their utility across different applications, making them indispensable in the toolkit of anyone venturing into AI.

Understanding and applying these methods requires a

blend of mathematical theory and computational practice. It's not just about plugging numbers into formulas; it's about understanding how those formulas come to be and how changing the inputs affects the outputs. This insight is crucial for tweaking AI models to perform better or diagnosing why they might be failing.

Numerical integration and differentiation are about more than just solving math problems. They are about applying mathematical concepts to real-world scenarios in AI, making informed decisions based on approximations, and continuously refining those decisions as more data becomes available or models evolve. This is the essence of mathematical application in artificial intelligence, blending theory with practice to solve complex problems.

Error Analysis

In the realm of artificial intelligence, numerical methods' precision can't be overstated. Error analysis, a critical component of these methods, helps us understand and mitigate the inaccuracies that inevitably arise during numerical computations. Let's explore why this is crucial for AI applications.

Firstly, numerical errors can broadly be classified into two types: truncation errors and round-off errors. Truncation errors occur when we approximate a mathematical process by terminating it after a finite number of steps. For instance, when we use a finite series to approximate a function, the difference between the actual function and the approximation is the truncation error. On the other hand, round-off errors stem from the limitations of representing numbers in computers. Since a computer can only handle a finite number of digits, rounding to the nearest representable number introduces these errors.

Understanding these errors is vital. For example, in a machine learning algorithm, minor errors in data processing can lead to significantly different outcomes. This sensitivity can be particularly problematic in high-stakes applications like autonomous driving or medical diagnosis systems, where precision is paramount.

To illustrate, let me share a personal experience. Once, while working on a machine learning model for predicting stock prices, a minor error in the numerical integration method used for calculating financial indicators led to a noticeable deviation in the prediction accuracy. This incident highlighted the cascading effects of seemingly trivial numerical errors and underscored the importance of rigorous error analysis.

To manage these errors, we employ various strategies. One common approach is to increase the precision of the numerical representation. However, this often comes at the cost of increased computational resources. Another strategy is to refine the algorithms themselves to minimize the introduction of errors.

In conclusion, error analysis is not just about identifying and correcting errors post-factum; it's about foreseeing potential inaccuracies and preemptively optimizing algorithms to handle them. This proactive approach in numerical methods ensures that AI systems are reliable and robust, capable of performing accurately in various applications.

Numerical Solutions of Equations

The ability to find numerical solutions to equations is crucial in artificial intelligence. This process allows AI systems to handle and interpret real-world data with precision. Let's explore how this is done, focusing on some key methods that are particularly useful in AI applications.

First up, we have the Newton-Raphson method. This technique is a powerhouse when it comes to solving nonlinear equations. Imagine you're trying to find the root of an equation (where the equation equals zero). Newton-Raphson uses tangents to converge on the correct solution iteratively. It starts with a guess and then refines this guess by evaluating the function and its derivative. The beauty of this method lies in its speed—converging to the root very quickly when the initial guess is close to the actual root.

Another essential method is the bisection method. This one is all about dividing and conquering. You start with an interval where the function changes sign (meaning the root must be within that interval). The bisection method zeroes in on the root by repeatedly bisecting this interval and selecting the subinterval where the sign change occurs. It's slower than Newton-Raphson but has the advantage of guaranteed convergence, making it a reliable choice when dealing with tricky functions.

For systems of linear equations, which are common in AI for modeling relationships and data fitting, iterative methods like the Jacobi method and the Gauss-Seidel method are invaluable. These methods approach the solution by iteratively improving guesses based on linear equations. The Jacobi method solves each equation for the desired variable and uses the old values for the other variables. On the other hand, Gauss-Seidel improves upon Jacobi by using the new values as soon as they are updated, often leading to faster convergence.

Each of these methods has its place in AI. Choosing the right one depends on the problem, the nature of the equations involved, and the required precision and speed of convergence. Understanding these methods empowers AI systems to perform at their best. It provides us with deeper insights into the mathematical structures underlying intelligent behavior.

So, whether it's optimizing a neural network, calibrating sensors in a robotics project, or modeling economic forecasts, numerical solutions to equations are at the heart of making AI systems function effectively in a complex and changing world.

Optimization Techniques

Optimization techniques are crucial in artificial intelligence, particularly when it comes to training models and solving complex problems where decisions need to be made about the best possible outcomes. These methods are not just about finding any solution but the best one according to a specific criterion, usually through minimizing or maximizing a function.

Let's start with the basics. At the heart of many AI systems, especially in machine learning, is a function measuring how well the system performs. This function is often called a cost, loss, or objective function. The goal of optimization is to tweak the system to minimize this function. The lower the value of the loss function, the better the performance of the AI system.

One of the simplest and most widely used methods for optimization is gradient descent. This technique involves looking at the function's gradient (essentially the slope of the function at a given point) to determine the direction in which the function decreases most rapidly. The function moves towards its minimum value by repeatedly taking steps in the opposite direction of the gradient.

However, gradient descent could be better and slower,

especially for complex functions. This has led to the development of more sophisticated methods like stochastic gradient descent, which uses a random subset of data to speed up the computations, and momentum-based methods, which help accelerate gradient vectors in the right directions, thus leading to faster converging.

Another popular method is the Newton-Raphson method, which uses an approach of finding zeros of a function derivative to find the minimum or maximum of the function. This method can be faster than gradient descent but requires the calculation of second derivatives, which can be computationally expensive.

For problems where derivatives are complex to compute or for discrete optimization problems, other techniques, such as genetic algorithms, simulated annealing, or particle swarm optimization, are used. Natural processes inspire these methods, which can be very effective for certain optimization problems.

In practice, choosing the proper optimization technique can depend on several factors, including the function's nature, the data size, the accuracy required, and the computational resources available. It's often a balance between precision and speed; sometimes, a combination of methods might be used to achieve the best results.

In summary, optimization is a fundamental concept in AI that helps improve models' decision-making capabilities. By understanding and applying the right optimization techniques, one can significantly enhance the performance of AI systems. Whether tuning a machine learning model or solving a complex scheduling problem, effective optimization is key to achieving high efficiency and performance.

Finite Element Analysis

Finite Element Analysis (FEA) is a powerful numerical method used extensively in engineering, physics, and, increasingly, artificial intelligence (AI) to solve complex structural and fluid dynamics problems. At its core, FEA breaks down a complicated problem into smaller, simpler parts that are easier to understand and solve. These smaller parts are called finite elements.

Imagine you're trying to understand the stress distribution in a bridge or the airflow around a high-speed train. FEA allows engineers and scientists to create a computational model of the structure or system, which is dissected into a mesh of smaller, discrete elements. Each element is considered to have uniform properties, and by solving the basic physics equations for each element, we can predict behaviors like deformation under stress or heat distribution in an engine.

The process begins with the creation of a geometric model, followed by the generation of a mesh. This meshing is critical as it defines how fine or coarse your simulation will be. A finer mesh generally provides more accurate results but requires more computational power and time. The elements in the mesh are interconnected at points called nodes, which play a crucial role in the analysis. The physical behaviors are calculated at these nodes, and the results are interpolated across the elements to provide a continuous picture of the phenomenon being studied.

The equations used in FEA are derived from fundamental principles, such as the conservation of mass, momentum, and energy. These equations are often too complex to solve analytically, especially for irregular shapes and boundary conditions. FEA provides a numerical solution to these equations, typically using methods like the Galerkin or the Rayleigh-Ritz method, which convert these equations into algebraic forms that can be solved using standard computational techniques.

One of FEA's key advantages is its versatility. It can be adapted to study various physical phenomena, including static (time-independent) problems, dynamic (time-dependent) problems, linear and nonlinear material behavior, fluid interactions, and coupled problems like thermo-mechanical issues where heat and force interact.

In the context of AI, FEA is particularly valuable for predictive modeling and simulations that feed into machine learning algorithms. For instance, in the design of robotic components, FEA can predict failure points or stress concentrations that can be used to train AI models for better decision-making in real-time operations.

Despite its numerous benefits, FEA has challenges.

The accuracy of the results heavily depends on the quality of the mesh and the assumptions and simplifications made in the model. Moreover, handling complex boundary conditions and nonlinear materials can complicate the analysis further.

In summary, Finite Element Analysis is a cornerstone technique in numerical methods. It provides detailed insights into complex systems that are critical for both traditional engineering fields and emerging areas in AI. Its ability to break down daunting physical problems into manageable elements makes it an indispensable tool in the arsenal of today's scientists and engineers.

Monte Carlo Methods

Monte Carlo methods are fascinating and powerful tools in numerical analysis, particularly useful in the fields of artificial intelligence and machine learning. These methods rely on random sampling to obtain numerical results, and they are typically used to solve problems that might be deterministic in principle but are too complex for analytical solutions.

Imagine you're trying to predict the outcome of a very complex system. Traditional algorithms might falter due to the sheer number of variables and interactions. This is where Monte Carlo methods come into play. By using randomness to sample from a probability distribution, these methods can offer approximations that get closer to the true value as more samples are taken.

One common application of Monte Carlo methods in AI is optimization problems, where you might want to find a global maximum or minimum of a function. Instead of exhaustively searching through all possible solutions, Monte Carlo methods allow you to explore the solution space randomly, which can be much more efficient for large datasets.

Another exciting application is in Bayesian inference, a statistical method in AI that updates the probability for a hypothesis as more evidence or information becomes available. Monte Carlo methods simulate the complex probabilistic models underpinning Bayesian approaches, helping to refine predictions and decisions in AI systems.

The beauty of Monte Carlo methods lies in their simplicity and flexibility. Given enough computational power, they can be applied to virtually any problem, making them indispensable tools in the arsenal of techniques used to tackle the complex mathematical problems encountered in AI.

Understanding and applying Monte Carlo methods will be crucial as we continue to push the boundaries of what AI can achieve. They offer a practical approach to solving high-dimensional problems and enhance our understanding of the stochastic nature of the world, which is often mirrored in the very data on which AI systems are trained.

Chapter Summary

- Numerical integration approximates the area under a curve by breaking it into smaller pieces, using methods like the Trapezoidal and Simpson's rules.
- Numerical differentiation estimates the rate of change by calculating the slope of secant lines using methods such as forward difference and central difference.
- These numerical methods are practical tools in AI for optimizing algorithms and processing data in fields like machine learning and computer vision.
- Error analysis in numerical methods identifies and mitigates inaccuracies, which is crucial for precision in AI applications like autonomous driving and medical diagnostics.
- Numerical solutions to equations, such as the Newton-Raphson and bisection methods, are essential for handling real-world data and optimizing AI systems.
- AI optimization techniques involve gradient descent and Newton-Raphson to minimize or maximize functions, improving model decision-making.
- Finite Element Analysis (FEA) breaks down complex problems into smaller parts, crucial

for engineering and AI applications like predictive modeling.

- Monte Carlo methods use random sampling to solve complex problems, useful in AI for optimization and Bayesian inference, enhancing the handling of high-dimensional data.

8

COMPLEX VARIABLES

Complex Numbers and Functions

At their core, complex numbers are an extension of the real numbers we are all familiar with. They are typically expressed in the form $a + bi$, where a and b are real numbers, and i is the imaginary unit with the property that $i^2 = -1$.

This might initially sound abstract, but think of it this way: if real numbers are a line, complex numbers are a plane. This two-dimensional nature allows them to capture dynamics that real numbers alone cannot, such as rotation and oscillation, which are crucial in understanding wave behaviors and other phenomena in physics and engineering.

Now, let's talk about the functions of complex variables. These functions take complex numbers as inputs and give complex numbers as outputs. They are fascinating because their behavior can be beautiful and bizarre. For instance, a function like $f(z) = z^2$, where z is a complex number not only squares the magnitude but also doubles the angle of z. This is a glimpse into how complex functions can transform the plane, leading to stunning visual representations in fractals and other complex geometric figures.

Moreover, complex functions are governed by rules that make them incredibly smooth and interconnected. The beauty of these functions lies in their differentiability. Being differentiable is a stronger condition in the complex world than in the real world. A differentiable complex function is smooth and conformal, meaning it preserves angles. This property is pivotal in many applications, including fluid dynamics and electrical engineering.

Understanding complex numbers and their functions opens up a new dimension of mathematical insight that is theoretical and immensely practical in technology and science. As we explore AI and its reliance on complex algorithms, recognizing the role of complex variables gives us a deeper appreciation of the mathematical underpinnings that make such technologies work. So, as we move forward, remember that these aren't just numbers on a page—they're the building blocks of the digital world around us.

Analytic Functions

An analytic function of a complex variable is defined primarily by the condition of being differentiable in a neighborhood of every point in its domain. This may sound a bit dense, but it's a powerful concept. Differentiability in this context means that the function is smooth and has a derivative at each point in its domain. Unlike real functions, for a function of a complex variable, being differentiable once implies that it is differentiable infinitely often and can be represented by a power series. This is known as the function's Taylor series expansion.

Why does this matter in AI? Many problems in AI, including those involving complex signal processing or the optimization of certain types of algorithms, can be modeled using analytic functions. The properties of analytic functions, such as their predictability and the ability to be expressed in series, make them particularly useful for these applications.

For instance, the behavior of analytic functions near essential singularities or their integral representations can often be leveraged to solve complex algorithm design problems or to analyze systems' stability. Moreover, the fact that the real and imaginary parts of an analytic function are not independent but linked by the Cauchy-Riemann equations helps provide constraints that simplify problems in two-dimensional space, a common scenario in AI applications.

Understanding analytic functions also prepares one to delve deeper into other complex topics, such as conformal mappings. These transformations preserve angles and are immensely useful in fields like computer graphics and robotics, areas where AI has significant applications.

In summary, the study of analytic functions equips AI practitioners with tools to handle complex mathematical models that appear in advanced AI systems, enhancing their ability to innovate and solve challenging problems.

Complex Integration

To start, let's understand the basic setup. In complex integration, we deal with functions of a complex variable and integrate these functions over a path or contour in the complex plane. This path or contour is not just a line but can be any curve connecting two points in the complex plane. The integral's value depends heavily on the path chosen, unlike in real integration, where the integral between two points is path-independent under certain conditions.

One of the most powerful results in complex integration is Cauchy's integral theorem. It states that if a function is analytic (complex differentiable) throughout a connected domain, then the integral of the function over any closed contour within that domain is zero. This theorem simplifies calculations and has many profound consequences in theoretical physics, engineering, and AI.

Why is this important for AI? In artificial intelligence, especially in fields like neural networks and machine learning, complex integration helps understand the behavior of complex-valued functions representing various physical, biological, and economic phenomena. For instance, the backpropagation algorithms used in training deep neural networks can be extended to complex-valued neural networks, where understanding complex derivatives and integrals becomes essential.

Moreover, the residues at poles of complex functions, calculated via the residue theorem (another result stemming from the principles of complex integration), are particularly useful in evaluating integrals in quantum mechanics, a field increasingly intertwined with AI for simulation and problem-solving.

In summary, complex integration is not just a theoretical exercise but a practical tool that enhances our ability to model, simulate, and ultimately understand complex systems in artificial intelligence. As we continue to push the boundaries of what AI can achieve, mathematical foundations like complex integration ensure that our algorithms are robust and mathematically sound.

Power Series and Residues

Diving into the world of complex variables, particularly the concepts of power series and residues can be exhilarating for anyone interested in the mathematical foundations necessary for artificial intelligence. Let's break these ideas down into simpler terms and explore their significance.

Starting with power series, think of them as an extension of polynomials. A power series is a sum of the powers of a variable. Still, unlike a polynomial, it can have infinitely many terms. For complex variables, a power series takes the form:

$$f(z) = a_0 + a_1(z - z_0) + a_2(z - z_0)^2 + a_3(z - z_0)^3 + \ldots$$

Here, z and z_0 are complex numbers, and a_n represents the series' coefficients. The beauty of power series in the context of complex functions is their ability to represent them over a circle in the complex plane, centered at z_0 and with a radius determined by the nearest point where the function misbehaves (like a singularity).

Why does this matter? In AI, understanding the behavior of functions, especially their smoothness and differentiability, is crucial for modeling and predictions. Power series provides a powerful tool for approximating complex functions, which can be used in various algorithms, including machine learning and neural networks.

Moving on to residues, these are closely tied to the concept of complex integration. The residue of a function at a point measures how the function behaves around that point. More formally, it's the coefficient of $1 / (z - z_0)$ in the function's Laurent series expansion around z_0.

Residues are particularly useful because they simplify the process of evaluating complex integrals thanks to the residue theorem. This theorem states that the integral of a function around a closed loop is $2\pi i$ times the sum of the residues of the function inside the loop. This powerful result allows for the evaluation of integrals that would otherwise be very difficult to compute directly.

In the realm of AI, these integrals often appear in the analysis of algorithms, especially those involving recursive relations or feedback loops. By understanding the residues, one can predict the behavior of these algorithms under various conditions, which is essential for stability analysis and optimization.

In summary, while power series and residues might initially seem abstract, their implications in AI are profound. They provide a deeper understanding of complex functions and equip us with tools to handle the intricacies of algorithms used in artificial intelligence.

Conformal Mapping

Conformal mapping is a fascinating concept in complex variables. It is instrumental in fluid dynamics, electromagnetism, and artificial intelligence. It involves a function that preserves angles. In simpler terms, if you have two curves crossing each other and apply a conformal map to these curves, they will still cross at the same angle, possibly at a new location or scale.

This property is crucial when dealing with complex functions because it helps maintain the data structure or the modeled physical phenomena, even under transformation. This makes conformal maps incredibly valuable for AI applications, especially in image processing and complex simulations where maintaining data's inherent structure is key.

Imagine you're working on an AI project that involves recognizing objects in images. By using conformal mapping, you can transform these images to make them easier for algorithms to process while preserving the crucial geometric relationships within the image. For instance, you could use conformal maps to standardize the size or orientation of objects in different images, making it easier for your AI to recognize these objects across varying scenarios.

Moreover, conformal mapping is about more than maintaining angles. It also has deep connections with analytic functions, which you can write down with power series in complex numbers. An analytic function can describe every conformal map between domains in the complex plane, assuming the function's derivative is non-zero. This ties back to the broader theme of complex integration and differentiability, which we explored in earlier sections.

Understanding and implementing conformal mapping can significantly enhance algorithm performance in practical AI applications, particularly in areas requiring detailed geometric accuracy. It's a powerful tool in the mathematician's toolkit, offering theoretical beauty and practical utility.

Applications of Complex Analysis

Complex analysis, a field that delves into the study of functions of complex variables, is more than just a theoretical endeavor. It has practical applications that span across various domains, including artificial intelligence (AI). Understanding these applications can provide significant insights into AI's theoretical and practical aspects.

One of the primary applications of complex analysis in AI is algorithm optimization. Complex functions, due to their differentiable nature, allow complex calculus to find optimum solutions efficiently. For instance, complex differentiation can be used to optimize functions in neural networks, enhancing their learning capabilities and efficiency.

Furthermore, complex analysis is crucial in signal processing and is foundational to many AI applications. The Fourier transform, a tool used to decompose functions into constituent frequencies, is based on complex numbers. It is extensively used in AI for processing audio signals, images, and other data forms, enabling the effective extraction and manipulation of useful features from large datasets.

Another significant application is in the field of control theory, which is vital for robotics and automated systems, areas where AI is heavily applied. The use of complex variables in control theory helps in designing controllers that can manage the behavior of dynamic systems, especially in complex environments where traditional methods might fail.

Quantum computing, an emerging field that promises to revolutionize computing using quantum phenomena, relies on complex analysis. Quantum algorithms, including those used for machine learning and optimization tasks, use complex numbers to represent states and perform computations exponentially faster than classical counterparts.

In fluid dynamics, complex analysis techniques help model fluid flows, which are essential in predictive models in AI for weather forecasting and simulating environmental scenarios. The mathematical foundation provided by complex analysis can be attributed to the potential to accurately predict and simulate complex fluid behaviors.

In conclusion, the applications of complex analysis in AI are vast and varied. From enhancing algorithm performance to enabling new computing paradigms, the contributions of complex variables prove to be fundamental in pushing the boundaries of what AI can achieve.

Chapter Summary

- Complex numbers are expressed as $a + bi$, where a and b are real numbers, and i is the imaginary unit with $i^2 = -1$.
- Complex numbers form a plane, representing dynamics like rotation and oscillation, which are crucial in physics and engineering.
- Functions of complex variables take and return complex numbers, with behaviors that include transformations like squaring magnitudes and doubling angles.
- Differentiable complex functions are smooth and conformal, preserving angles, which is essential in fields like fluid dynamics and electrical engineering.
- Analytic functions, or holomorphic functions, are complex functions that are infinitely

differentiable and can be expressed as power series. They are useful in AI for modeling and optimization.

- Complex integration involves integrating functions over paths in the complex plane with applications in AI such as neural networks and machine learning.
- Power series represent complex functions as infinite sums, which are helpful for approximating functions in AI algorithms. At the same time, residues help evaluate complex integrals, aiding in algorithm analysis.
- Conformal mapping preserves angles between curves under transformation, which is beneficial in AI for tasks like image processing and maintaining geometric relationships in data.

9

DIFFERENTIAL EQUATIONS

First Order Differential Equations

Diving into the world of first-order differential equations, we find ourselves at the intersection of mathematics and real-world applications. These equations are not just abstract concepts; they are tools that help us model everything from population growth to electrical circuits.

A first-order differential equation involves derivatives of the first degree. Simply put, it relates a function, its derivative, and the independent variable. The general form can be expressed as:

$$dy/dx = f(x, y)$$

Here, y is a function of x, and f is some function that connects x and y.

One of the simplest and most illustrative examples of a first-order differential equation is the exponential growth model, described by the equation:

$$dy/dx = ky$$

Here, k is a constant. This model is widely used to describe phenomena such as unchecked population growth or the spread of a virus, where the rate of increase of the population (or infected individuals) is proportional to the current population.

Solving these equations often involves finding an integrating factor or separating variables, transforming the equation into a more manageable form. For instance, with the separation of variables, we can rearrange $dy/dx = ky$ to $dy/y = k\,dx$.

This can then be integrated on both sides to give: $ln\ |y| = kx + C$ where C is the integration constant. Exponentiating both sides, we find $y = Ce^{(kx)}$. This reveals how the function y evolves over time.

In the context of artificial intelligence, understanding these equations is crucial when we deal with systems that change dynamically over time. For example, in neural networks, differential equations can model how input changes affect output changes, crucial for training and optimization processes.

Moreover, the advent of neural ordinary differential equations presents an exciting frontier in which these mathematical constructs are not just tools for analysis but integral components of the learning model itself.

In summary, first-order differential equations are foundational in understanding and designing artificial intelligence systems. They provide a framework for modeling dynamic changes and offer insights critical to developing robust AI systems.

Second Order Linear Equations

Diving into the world of second-order linear differential equations, we find ourselves at the heart of many dynamic systems crucial for AI applications. These equations are typically expressed in the form:

$$a(x)y'' + b(x)y' + c(x)y = f(x)$$

Here, y'' denotes the second derivative of y with respect to x. These equations are fundamental in modeling phenomena where the current state depends not only on its immediate past but also on the rate at which it's changing.

To understand these equations better, let's break down their components. The functions $a(x)$, $b(x)$, and $c(x)$ are coefficients that can vary with x, adding a layer of complexity and realism to the model. The function $f(x)$ on the right side of the equation represents an external force or input affecting the system, which can be zero or any other function.

One of the most enlightening examples of a second-order linear differential equation is the simple harmonic oscillator, described by the equation:

$$y'' + \omega^2 y = 0$$

Here, ω represents the angular frequency of the oscillator. This model is pivotal in physics for studying systems like springs and pendulums, but it also appears in AI in areas like signal processing and system control.

The general solution to these equations combines a homogeneous solution (solving $a(x)y'' + b(x)y' + c(x)y = 0$) with a particular solution, which is derived from the non-homogeneous equation. The homogeneous solution involves characteristic equations and possibly complex numbers, depending on the discriminant $b^2 - 4ac$. The particular solution, on the other hand, can often be found using methods like undetermined coefficients or variation of parameters, depending on the nature of $f(x)$.

In AI, understanding and solving these equations enable the modeling of systems where future state predictions are based on current and past data. For instance, in robotics, the motion of a robot arm can be modeled using second-order differential equations, where the arm's position, speed, and acceleration are all taken into account to ensure smooth and precise movements.

Moreover, in machine learning, especially in areas dealing with time series prediction or signal processing, these equations help design algorithms that can predict future events based on past observations, which is crucial for tasks like stock market prediction or weather forecasting.

In summary, second-order linear differential equations are not just mathematical constructs but powerful tools in the AI toolkit. They help bridge the gap between dynamic real-world phenomena and artificial intelligence's predictive power. Understanding their structure and solutions enables AI practitioners to build more robust and accurate models for various applications.

Systems of Differential Equations

When we explore differential equations in the context of artificial intelligence, systems of differential equations stand out as particularly crucial. These systems are not just a single equation but a set working together, often describing multiple interrelated phenomena. This interconnectedness makes them valuable in modeling complex real-world processes, which is precisely what we aim to understand and predict in AI.

Imagine you're trying to model the weather. One equation might represent temperature, another humidity, and yet another the pressure. Each of these factors influences the others, and their relationships can be expressed and explored through systems of differential equations. This is similar to how various features and factors interact in many AI models, especially in dynamic environments.

Differential equation systems can be classified mainly into two types: linear and nonlinear systems. Linear systems are generally easier to solve and analyze when the equations involve only linear terms of the variables and their derivatives. Nonlinear systems, however, involve terms that are nonlinear functions of the variables and their derivatives, making them more complex and more capable of capturing the intricacies of real-world dynamics.

We often use methods like matrix algebra, which you recall from the Linear Algebra chapter, to solve these systems. We can represent the system in a matrix form and then apply various algebraic techniques to find solutions. This approach is efficient and aligns well with computational methods used in AI, where handling large matrices is commonplace.

Moreover, numerical methods, discussed in the Numerical Methods chapter, play a pivotal role when analytical solutions to these systems are not feasible. Techniques such as Euler's method, Runge-Kutta method, and others allow us to approximate solutions with great accuracy, which is often sufficient for practical purposes in AI applications.

Understanding and applying systems of differential

equations enable AI not just to predict outcomes but also to understand how different variables influence each other over time. This is essential in fields like robotics, where the interaction of various mechanical parts must be precisely coordinated, or in economics, where multiple factors influence market trends.

In summary, systems of differential equations are not just mathematical constructs but powerful tools in the arsenal of AI. They provide the framework to model, analyze, and predict complex dynamic systems in an interconnected world.

Laplace Transforms

Laplace transforms are a powerful mathematical tool, especially when solving differential equations crucial in modeling various AI applications. Imagine you're dealing with a system where you need to predict future events based on known rates of change—this is where differential equations come into play. Now, add Laplace transforms to your toolkit, and you have a method to simplify many of these complex equations into a more manageable form.

So, what exactly is a Laplace transform? The technique transforms a time-domain function into a complex frequency-domain function. This sounds abstract but think of it as converting a time-based puzzle into a frequency-based puzzle, which often turns out to be easier to solve.

Why are they so valuable for AI? In artificial intelligence, we often deal with systems that change over time, whether predicting stock prices, understanding speech, or controlling a robot. Differential equations can describe these systems. Applying the Laplace transform, these equations, which can be notoriously difficult to solve directly, are transformed into algebraic equations, which are typically much more straightforward.

Here's a basic rundown of how it works: The Laplace transform takes a function of time, $f(t)$, and transforms it into a function of a complex variable s, denoted as $F(s)$. This function $F(s)$ encapsulates all the information of the original function but in a form that often simplifies the operations needed to analyze or design a system.

For instance, consider a differential equation describing a machine part's cooling process in a robotic assembly. The rate of cooling at any time t might be given by a differential equation. By applying the Laplace transform, this time-dependent equation is re-framed in terms of s, which can then be manipulated algebraically to solve for $F(s)$ and eventually transformed back to give the temperature at any time t.

Moreover, Laplace transforms are about more than just making equations easier to solve. They also help in understanding the behavior of systems. For example, the poles and zeros of the transformed function $F(s)$ can tell us about the stability of the system, its responsiveness, and how it behaves over time without solving the differential equation explicitly.

In summary, Laplace transforms a bridge between the complex dynamic behaviors of systems critical in AI and the static, algebraic methods that are easier to analyze and understand. They allow AI practitioners to focus on designing and improving systems rather than getting bogged down by complex calculus, making them an indispensable part of the AI mathematician's toolbox.

Fourier Series and PDEs

Diving into the world of Fourier series and partial differential equations (PDEs) opens up a fascinating chapter in the mathematical playbook, especially regarding applications in artificial intelligence. Let's break down these concepts into digestible parts to see how they play pivotal roles in modeling and solving complex problems that AI systems often face.

Starting with the Fourier series, think of it as a mathematical tool that decomposes any periodic function or signal into the sum of a (possibly infinite) set of simple oscillating functions, namely sines and cosines. The beauty of this lies in its ability to transform complex problems in the time domain into much simpler forms in the frequency domain. For AI, this is particularly useful in tasks involving signal processing, image analysis, and even in advanced algorithms for learning periodic patterns in data.

Now, let's talk about partial differential equations (PDEs). PDEs are used to formulate problems involving functions of several variables and are either solved by hand or used to create a computer model. In AI, PDEs are crucial for simulating physical processes and optimizing systems, and even in sophisticated neural networks like those used in deep learning to understand spatial-temporal dynamics.

By combining the Fourier series with PDEs, we can tackle many problems. For instance, Fourier methods can help filter and smooth data in image processing, a preliminary step in edge detection algorithms crucial for object recognition tasks. Moreover, the heat equation, a type of PDE, can be solved using the Fourier series to model how different materials conduct heat over time—a process analogous to understanding how information spreads in a network.

Understanding these mathematical tools enhances the capability of AI systems and provides a robust framework for innovation. By mastering the Fourier series and PDEs, AI practitioners can design more efficient algorithms that can learn from complex datasets, predict future trends, and even solve problems that were once thought to be intractable.

Applications to Dynamic Systems

Differential equations are not just mathematical expressions; they are the language of change and dynamics in the universe. They help us model everything from the most straightforward mechanical systems to the most complex behaviors in financial markets. Differential equations are indispensable for describing how systems evolve over time, particularly in dynamic systems.

Consider a dynamic system like a swinging pendulum. At first glance, it's a simple back-and-forth motion governed by gravity and momentum. However, to predict its exact position at any given moment, we must solve a differential equation that accounts for angles, gravity, air resistance, and initial force. This second-order linear differential equation helps us understand not just pendulums but any system with similar dynamics, such as specific electrical circuits or even the motion of planets in a simplified model.

In artificial intelligence, dynamic systems modeled by differential equations are crucial for developing algorithms that can predict and adapt to changing environments. For instance, when programming a drone to navigate varying terrains autonomously, differential equations model the changing conditions and guide the drone's responses in real-time.

Thus, understanding differential equations opens up a world of possibilities in AI. They are not merely academic; they are tools that, when wielded with skill, can lead to innovations in robotics, economics, environmental modeling, and beyond.

Chapter Summary

- First-order differential equations are used to model real-world phenomena like population growth and electrical circuits, with a general form expressed as $dy/dx = f(x, y)$.
- Techniques such as integrating factors and separation of variables are used to solve these equations, exemplified by the exponential growth model $dy/dx = ky$.
- Second-order linear differential equations, expressed as $a(x)y'' + b(x)y' + c(x)y = f(x)$, model dynamic systems where the current state depends on its immediate past and the rate of change.
- These equations are crucial in AI for modeling systems like robotic arms and for tasks such as stock market prediction, using solutions that combine homogeneous and particular solutions.
- Systems of differential equations, both linear and nonlinear, model complex, interrelated phenomena and are solved using methods like matrix algebra and numerical techniques.

- Laplace transforms simplify differential equations by converting time-domain functions into frequency-domain functions, aiding in system design and analysis in AI.
- Fourier series and partial differential equations (PDEs) are used in AI for tasks like signal processing and image analysis, transforming complex time-domain problems into simpler frequency-domain problems.
- Differential equations describe dynamic systems and are essential in AI for modeling and predicting behaviors in environments like autonomous drone navigation.

10

OPTIMIZATION TECHNIQUES

Linear Programming

Linear programming is a powerful mathematical method to optimize a linear objective function subject to linear inequality or equality constraints. This technique has vast applications in various fields, including business, economics, and engineering, particularly regarding resource allocation, production planning, and scheduling.

At its core, linear programming involves determining a linear objective function's maximum or minimum value. This function typically represents some quantity we want to maximize or minimize, such as profit, cost, or time. The constraints, on the other hand, represent the limitations or requirements of the problem, such as resource availability or demand satisfaction.

To solve a linear programming problem, various

methods can be used, the most famous of which is the Simplex algorithm. Developed by George Dantzig in 1947, the Simplex algorithm involves a series of steps to move from one vertex of the feasible region to another, improving the value of the objective function at each step until the maximum or minimum value is reached.

Another approach is to use graphical methods for problems involving two variables. By graphing the constraints, one can visually identify the feasible region and then determine the optimal solution by evaluating the objective function at the vertices of this region.

Linear programming also extends to more complex scenarios through its variations, such as integer linear programming (ILP), where some or all of the decision variables are constrained to be integers, and mixed-integer linear programming (MILP), which involves both integer and continuous variables.

In artificial intelligence, linear programming can benefit optimization problems where decisions must be made under certainty. AI systems can leverage linear programming to automate decision-making processes, optimize operational efficiencies, and predict future outcomes based on constraints and objectives.

Understanding linear programming provides a solid foundation for dealing with more complex, non-linear optimization problems that one may encounter in advanced AI applications. This makes it an essential tool in the arsenal of techniques available to AI practitioners aiming to solve real-world problems efficiently.

Non-linear Optimization

Diving into the world of non-linear optimization, we're stepping into a realm where the straightforward paths of linear programming no longer apply. This is the territory where the functions we aim to optimize (or minimize) are not straight lines but curves, which can be as unpredictable as a roller coaster track.

Non-linear optimization is crucial because it mirrors the complexity of real-world problems. From adjusting parameters in a machine learning model to minimize error to finding the most efficient design in engineering, these problems require us to navigate through multiple peaks and valleys to search for an optimal solution.

The core challenge here is that non-linear functions can have multiple local minima and maxima, making it tricky to find the global optimum. Imagine hiking in a range filled with numerous hills and valleys; the highest peak is only sometimes visible from the foot of any hill. Similarly, in non-linear optimization, simple descent methods that move towards the nearest low point can get stuck in a local minimum, far from the best possible solution.

Various sophisticated algorithms have been developed to tackle these challenges. One popular method is the

Newton-Raphson technique, which uses derivatives to find where the function's slope hits zero—a potential minimum or maximum. However, this method assumes the function is differentiable and can run into trouble if it's not.

Another powerful strategy is gradient descent, which iteratively adjusts the function's variables to move towards the steepest descent, ideally leading to a global minimum; for problems where the function's landscape is particularly rugged, simulated annealing or genetic algorithms might be employed. These methods introduce randomness into the search process, helping to escape local minima that could trap more straightforward approaches.

In the context of AI, non-linear optimization becomes even more significant. Consider neural networks, which learn by adjusting weights to minimize a loss function. Depending on the network's complexity and the data, this function is typically non-linear. Efficiently finding a good set of weights is crucial for the training process, directly impacting the performance of the AI system.

In summary, non-linear optimization finds the best possible outcomes in a world where straight paths are the exception rather than the rule. It's a fascinating area that combines deep mathematical theory with practical techniques that help solve some of the most complex problems in science, engineering, and economics.

Convex Optimization

In artificial intelligence, the ability to find optimal solutions efficiently is crucial, and this is where convex optimization comes into play. It's a subset of optimization that deals specifically with convex functions, where the line segment between any two points on the function's graph does not lie below the graph itself. This characteristic leads to some powerful advantages in solving optimization problems.

Firstly, convex optimization problems are inherently simpler because any local minimum is also a global minimum; there are no tricky dips and peaks like in non-convex functions. This means that the solutions are easier to find and more predictable and reliable, which is a big deal when you're dealing with AI algorithms that need to perform well consistently.

The applications of convex optimization in AI are vast. For instance, many machine learning models, including support vector machines and logistic regression, rely on convex optimization to determine the best fit for the model parameters. This process involves minimizing a cost function that measures how well the model predicts the desired outcomes. By leveraging convex optimization, these models can be trained more efficiently and with better outcomes.

Moreover, convex optimization is more than finding a function's minimum. It's also about feasibility and resource allocation, integral to operations research and many AI applications. For example, in network flow problems, convex optimization can help maximize throughput or minimize energy use in a system, ensuring optimal resource use.

Understanding convex optimization requires a good grasp of derivatives and subgradients, as these mathematical tools are used to navigate the function's surface and move toward the minimum point. It also often involves working with constraints, which are conditions that the solution must satisfy. These constraints can define boundaries and relationships in the optimization problem, adding an extra layer of complexity.

Gradient Descent Methods

Gradient descent is a fundamental optimization algorithm machine learning uses to minimize a function. Picture this: you're in a thick, foggy valley and must find the lowest point. You can't see far ahead due to the fog, but you can feel the slope of the ground beneath your feet. You eventually reach the valley's bottom by consistently moving in the direction that slopes downward. This is how gradient descent works.

The method involves taking the function's gradient (or the slope) at a given point and moving in the direction that results in the steepest descent. The learning rate determines the size of the steps taken in the search for the minimum. This crucial parameter needs to be set carefully. A lower learning rate makes the descent painfully slow. At the same time, a too-large rate can lead to overshooting the minimum, possibly diverging from the solution.

In practice, gradient descent starts with an initial guess. It iteratively updates this guess by moving toward the negative gradient. The updates continue until the changes are infinitesimally small, indicating that the minimum has likely been reached or until a set number of iterations are completed.

There are several variants of gradient descent, each suited to different types of problems:

1. **Batch Gradient Descent:** Computes the gradient using the entire dataset. This is precise but can be slow and computationally expensive with large datasets.
2. **Stochastic Gradient Descent (SGD):** This method computes the gradient using a single sample at each iteration. It is faster and can help escape local minima, but the path to convergence can be noisy.
3. **Mini-batch Gradient Descent:** This variant strikes a balance by computing the gradient against small batches of data. It is often

preferred in practice due to its efficiency and relatively smooth convergence.

Gradient descent is powerful but has limitations. It's prone to getting stuck in local minima instead of finding the global minimum. This is particularly problematic in non-convex functions common in deep learning. Additionally, its performance is highly sensitive to the choice of the learning rate and the initial starting point.

Despite these challenges, understanding and implementing gradient descent is a crucial skill in the toolkit of anyone venturing into AI and machine learning. It provides a foundational building block for many advanced optimization algorithms used across various applications in the field.

Stochastic Methods

In the realm of optimization techniques, stochastic methods stand out for their unique approach to finding solutions by incorporating randomness into the process. Unlike deterministic methods, which follow a fixed path to seek the optimum, stochastic methods introduce an element of chance, making them particularly useful in scenarios where the landscape is rugged, or the objective function is noisy.

One of the most popular stochastic methods is Simulated Annealing. Inspired by the process of annealing in metallurgy, this technique involves heating and slowly cooling a material to decrease defects. In optimization, this metaphor translates to exploring the solution space by allowing occasional uphill moves, avoiding local minima, and aiming for a global optimum as the system cools.

Another key player in stochastic methods is the Genetic Algorithm. This approach mimics natural selection, where the fittest individuals are chosen for reproduction to produce offspring of the next generation. In mathematical terms, solutions from one iteration are used to form a new pool of solutions, and this process is repeated until the best solution is found. It's particularly effective for solving problems where the solution space is vast and poorly needed to be better understood.

Stochastic Gradient Descent (SGD) offers a more nuanced approach. It modifies the traditional gradient descent algorithm, which adjusts parameters in the opposite direction of the gradient of the objective function. Instead of calculating the exact gradient, SGD estimates it using a randomly selected subset of data. This randomness helps to speed up computation significantly, especially in large-scale machine-learning problems.

Each method has its strengths and is suited to particular optimization problems. Simulated annealing is robust against getting trapped in local minima, making it suitable for complex landscapes. Genetic Algorithms are excellent for problems where the solution space is discrete, and the global structure is unknown. Meanwhile, Stochastic Gradient Descent shines in large-scale machine learning where the data set is too large to compute exact gradients efficiently.

Incorporating randomness might seem counterintuitive when precision is typically the goal in mathematical computations. However, as we've seen, the stochastic approach provides a powerful toolkit for tackling optimization problems that might otherwise be intractable. By embracing uncertainty and variability, these methods open up new possibilities for finding optimal solutions in complex and dynamic environments.

Optimization in Machine Learning

In machine learning, optimization is the backbone that helps fine-tune models to achieve the best possible performance. At its core, optimization revolves around adjusting the parameters of models to minimize or maximize a particular function. This function, often called the "loss function" or "objective function," measures how well the model's predictions align with the actual data.

One of the most popular methods used in this context is gradient descent. This technique involves taking small, iterative steps in the direction that most steeply reduces the loss. It's akin to descending a hill in the fog; you can't see the bottom, so you take steps in the steepest direction at each moment. In mathematical terms, this involves calculating the loss function's gradient (or derivative) concerning the model parameters and then updating the parameters in the opposite direction of the gradient.

However, gradient descent isn't a one-size-fits-all solution. There are several variants, each suited to different kinds of problems. For instance, stochastic gradient descent (SGD) updates parameters using only a single data point at a time, which makes it faster and less memory-intensive for large datasets. On the other hand, batch gradient descent uses the entire dataset to compute the gradient, which can be more stable but requires more computational resources.

Another critical aspect of optimization in machine learning is overfitting. This occurs when a model learns the underlying pattern and noise in the training data, leading to poor performance on new, unseen data. Techniques like regularization are used to prevent this. Regularization methods add a penalty term to the loss function to discourage the model from becoming overly complex.

Moreover, the optimizer's choice can significantly impact the training speed and quality of the final model. Beyond SGD, advanced optimizers like Adam and RMSprop have been developed to adjust the learning rate dynamically during training. These methods often lead to faster convergence and can handle non-stationary objectives and noisy problem settings.

In summary, optimization in machine learning is a dynamic field that combines deep mathematical concepts with practical strategies to train models effectively. It's not just about finding the best parameters but also about generalizing the model well to new data, remaining computationally feasible, and converging to a solution in a reasonable amount of time. As machine learning continues to evolve, so too will the optimization techniques, continually pushing the boundaries of what's possible in AI applications.

Chapter Summary

- Linear programming optimizes a linear objective function with linear constraints. It is widely used in resource allocation and scheduling in business, economics, and engineering.
- The Simplex algorithm, developed in 1947 by George Dantzig, is a crucial method in linear programming. It moves through the vertices of the feasible region to optimize the objective function.
- Linear programming variations include integer linear programming (ILP) and mixed-integer linear programming (MILP), which involve integer and continuous variables.
- Non-linear optimization deals with curved

functions, not straight lines, making it essential for complex real-world problems like machine learning model adjustments.

- Non-linear functions often have multiple local minima and maxima, requiring sophisticated algorithms like Newton-Raphson and gradient descent to find the global optimum.
- Convex optimization, a subset of optimization, is crucial in AI because it is simple to solve problems where any local minimum is also a global minimum.
- Gradient descent is a fundamental optimization method in machine learning. It uses the function's slope to move iteratively towards the minimum.
- Stochastic methods like Simulated Annealing and Genetic Algorithms incorporate randomness to effectively navigate complex solution spaces and avoid local minima.

11

STATISTICS FOR MACHINE LEARNING

Descriptive Statistics

Descriptive statistics are the cornerstone of data analysis in machine learning, providing a powerful way to summarize and understand complex datasets. At its core, descriptive statistics help to describe and show the features of a specific dataset by obtaining short summaries about the sample and measures of the data.

The main components of descriptive statistics include measures of central tendency, variability, and distribution shape. Measures of central tendency include the mean, median, and mode, which help us understand the data's average behavior or central location. For instance, the mean provides an arithmetic average, the median gives the middle value when data is ordered, and the mode represents the most frequently occurring value in the dataset.

Variability, or dispersion, tells us about the spread of the data. Key measures include the range, variance, standard deviation, and interquartile range. The range is the difference between the highest and lowest values, giving a quick sense of the spread. Variance and standard deviation, on the other hand, provide insights into how much the data deviates from the mean. The interquartile range, which measures the spread of the middle 50% of the data, helps understand the variability while being resistant to outliers.

Another aspect of descriptive statistics is the shape of the data's distribution, which can be described using skewness and kurtosis. Skewness measures the asymmetry of the data distribution, indicating whether the data points fall to the left or right of the mean. Compared to a normal distribution, kurtosis tells us about the distribution's heavy or light tails.

In machine learning, these statistical summaries allow us to perform initial diagnostics and make informed decisions about the models. For example, understanding the distribution of data can help choose the right algorithm or transform data to meet model assumptions. Moreover, we can preprocess datasets by identifying outliers, skewness, and other anomalies to improve model accuracy.

Descriptive statistics also play a crucial role in exploratory data analysis, where visual tools like histograms, box plots, and scatter plots are used alongside numerical measures to uncover patterns, trends, and relationships in the data.

In summary, descriptive statistics provide a fundamental first step in data analysis, offering a quick and insightful overview of the data's main characteristics. This not only aids in understanding the data but also guides the subsequent steps in the machine-learning workflow, including data preprocessing and model selection.

Inferential Statistics

At its core, inferential statistics deals with probabilities. It allows us to say, with some confidence, how likely it is that the conclusions we draw from our data samples are true for the entire population. This is particularly useful in machine learning, where we often work with data samples rather than entire populations.

One of the most common applications of inferential statistics in machine learning is hypothesis testing. This involves making an assumption (the hypothesis) about a dataset and then determining whether the data supports this hypothesis. For instance, if we're developing a model that predicts whether an email is spam, we might hypothesize that specific keywords significantly increase the likelihood of an email being spam. We can use sample data to statistically support or refute our assumption through hypothesis testing.

Another critical tool in inferential statistics is the confidence interval. This gives a range of values for an unknown parameter (e.g., the mean or proportion) with a certain degree of confidence. For example, we might be 95% confident that a user's average time on a website is between 5 and 7 minutes. This interval helps make decisions that are not just based on sample data but indicative of the larger population.

Regression analysis, another inferential technique, allows us to understand the relationship between variables and how they contribute to our study outcome. In machine learning, this is often used in predictive modeling. For example, regression could help us understand how age and income predict a person's likelihood of purchasing a product.

Lastly, inferential statistics also involves the use of Bayesian inference, a method of statistical inference in which Bayes' theorem is used to update the probability for a hypothesis as more evidence or information becomes available. This is incredibly powerful in machine learning for developing algorithms that adaptively improve as they are exposed to more data over time.

Understanding and applying inferential statistics is essential for anyone working in machine learning. It provides the mathematical foundation for making precise decisions and predictions from imperfect or incomplete information.

Predictive Modeling

Predictive modeling is a statistical technique that harnesses mathematics and statistics to predict outcomes. At its core, predictive modeling uses historical data to build a mathematical solution that can forecast future outcomes with significant reliability. In machine learning, predictive modeling is often synonymous with supervised learning.

Now, let's break down the process. Initially, you collect historical data, often termed training data in machine learning. This data must be relevant to the problem you are trying to solve. For instance, if you want to predict housing prices, your data might include features like square footage, number of bedrooms, and house age.

Once you have this data, the next step is to create a model. This involves selecting an algorithm to turn your data into a predictive tool. The choice of algorithm depends on the type of data you have and the prediction you need to make. Common algorithms include linear regression, logistic regression, decision trees, and neural networks.

After selecting your algorithm, you train your model by running your algorithm on the data. During this phase, the algorithm will learn by adjusting its parameters. It tries to find patterns in the training data that lead to the outcomes you're interested in predicting.

Once the model is trained, you can use it to make predictions. You do this by feeding new data into the model and letting the model apply what it has learned to this new data. The output is a prediction about the likely outcome.

However, it's crucial to measure how accurate your model is. This is typically done through validation, where you test your model using a new data set it hasn't seen before. This helps to ensure that your model can make accurate predictions on new, unseen data, not just the data it was trained on.

Predictive modeling is powerful because it allows you to make informed decisions about the future. It's used in a variety of fields, from finance to healthcare, to predict everything from stock prices to patient outcomes. In each case, the better your model, the more accurate your predictions will be, allowing for smarter, data-driven decisions.

In summary, predictive modeling is a fundamental aspect of machine learning that involves using historical data to predict future outcomes. Understanding and applying thc right statistical techniques and algorithms can unlock valuable insights hidden in your data.

Machine Learning Algorithms

Diving into the world of machine learning algorithms, it's crucial to understand how they are intertwined with statistical methods to extract meaningful patterns and insights from data. Machine learning utilizes algorithms to parse data, learn from it, and then decide or predict something in the world. These algorithms are fundamentally statistical models that fit the data.

Firstly, let's talk about supervised learning algorithms. These include popular methods like linear regression, logistic regression, and support vector machines. We predict a continuous output variable in linear regression based on the input variables. It's like fitting a line in a two-dimensional space or a plane in a three-dimensional space that best approximates all the data points. Logistic regression, on the other hand, deals with categorical output variables and is widely used for classification tasks, such as determining whether an email is spam.

Then, there's unsupervised learning, where the data isn't labeled, and the algorithm tries to learn the underlying patterns without explicit instructions on what to look for. Clustering is a prime example of unsupervised learning, with k-means clustering being particularly popular. This algorithm partitions the data into k distinct clusters based on feature similarity.

Another critical area is ensemble methods, which combine multiple machine learning models to produce better predictive performance than could be obtained from any of the individual models alone. Algorithms like Random Forests and Gradient Boosting Machines fall into this category. Random Forests, for instance, build multiple decision trees and merge them to get a more accurate and stable prediction.

Deep learning, a subset of machine learning, uses algorithms inspired by the structure and function of the brain called artificial neural networks. It is particularly useful in processing and learning from large amounts of data and excels in tasks like image recognition, natural language processing, and speech recognition.

Each of these algorithms requires a solid statistical foundation to ensure accurate data handling, effective model training, and the correct interpretation of results. Understanding the distribution of data, hypothesis testing, and confidence intervals helps fine-tune models and decide which algorithms work best for a particular problem.

In summary, machine learning isn't just about coding algorithms—it's about understanding data through a statistical lens. This statistical foundation helps build robust models and interpret their outputs, ensuring that decisions made based on machine learning are as informed and reliable as possible.

Model Evaluation

In machine learning, evaluating a model's performance is not just a step in the process—it's a cornerstone of successful implementation. This evaluation phase helps us understand how well our model will likely perform when encountering new, unseen data. It's all about ensuring reliability, efficiency, and accuracy in predictive analytics.

Let's summarize the key methods and metrics used to evaluate machine learning models. First up, we have the confusion matrix, a handy tool that helps us visualize the performance of our classification model. It shows the number of correct and incorrect predictions broken down by each class. This leads us directly into discussing accuracy, precision, recall, and the F1 score—metrics that give us a clearer picture of our model's performance beyond just a simple accuracy percentage.

Accuracy might tell you the overall correctness of the model. Still, precision and recall will let you know where it's getting things right—and crucially, where it's going wrong. Precision focuses on the relevancy of the model's predictions. At the same time, recall deals with how well the model can pick up on all relevant instances. The F1 score harmonizes these two metrics, providing a score that balances precision and recall, which is particularly useful when dealing with imbalanced datasets.

Moving beyond classification, in regression tasks, we often rely on metrics like Mean Absolute Error (MAE), Mean Squared Error (MSE), and Root Mean Squared Error (RMSE) to quantify the difference between the values predicted by the model and the actual values. These metrics help understand the magnitude of error in predictions. RMSE is particularly sensitive to large errors due to its squaring of the residuals.

Another crucial aspect of model evaluation is training and testing datasets. The general practice is to train our models on a large chunk of the available data and then test them on a separate set that the model has yet to see. This approach helps mitigate the risk of overfitting, where a model might perform exceptionally well on training data but poorly on any new data.

Cross-validation is another robust method for assessing a model's effectiveness. It involves dividing the dataset into several subsets and training the model multiple times, each time with a different subset held out for testing. This method provides a more comprehensive insight into how the model will perform across different subsets of data.

Lastly, it's essential to consider the ROC curve and AUC—Area Under the Curve—when evaluating classification models. These tools help understand the trade-offs between true positive rates and false positive rates at various threshold settings, providing a holistic view of model performance.

In summary, model evaluation in machine learning isn't just about picking the right metrics but also about understanding the deeper insights they provide about your model's capabilities and limitations. It's a critical step that bridges the gap between theoretical development and practical, real-world application, ensuring that our models are accurate, reliable, and robust.

Advanced Statistical Techniques

In machine learning, harnessing advanced statistical techniques can significantly enhance model performance and provide deeper insights into data. These techniques not only refine the predictions but also help understand the complex relationships hidden within the data.

One such advanced method is multivariate analysis, which involves examining multiple variables to understand their relationships and influence on each other. This is particularly useful in scenarios where variables are interconnected, and simple univariate analysis might miss critical insights.

Another pivotal technique is the use of hierarchical models, especially in situations where data is nested or grouped. These models are adept at handling data with multiple levels of hierarchy, allowing for more precise conclusions by considering the variance within each group and between groups.

Machine learning also greatly benefits from the application of time series analysis, especially when dealing with sequential data, such as stock prices or weather patterns. Techniques like ARIMA (AutoRegressive Integrated Moving Average) and seasonal decomposition provide frameworks for forecasting future values based on previously observed data, incorporating trends and cyclicality.

Moreover, survival analysis presents another layer of complexity, offering tools to address questions about durations until an event of interest occurs. This is crucial in fields like medicine or customer churn analysis, where understanding the 'when' can be as important as the 'why.'

Lastly, integrating Bayesian statistics into machine learning offers a robust way to deal with uncertainty and incorporate prior knowledge into the models. Bayesian methods update the probabilities as more evidence becomes available, making them incredibly powerful in predictive analytics.

Data scientists can build more robust, efficient, and insightful models by leveraging these advanced statistical techniques. This pushes the boundaries of what machines can learn and opens up new avenues for innovation and discovery in artificial intelligence.

Chapter Summary

- Descriptive statistics summarize complex datasets in machine learning, focusing on central tendency, variability, and distribution shape measures.
- Inferential statistics in machine learning involve hypothesis testing, confidence intervals, and regression analysis to predict larger populations from sample data.
- Predictive modeling uses historical data to forecast future outcomes, employing algorithms like linear regression and neural networks, and is validated through accuracy testing.
- Machine learning algorithms range from supervised learning (e.g., linear regression) to unsupervised learning (e.g., clustering) and are essential for data pattern recognition.
- Model evaluation in machine learning uses metrics like accuracy, precision, recall, and RMSE, as well as methods like cross-validation, to ensure models perform well on new data.
- Advanced statistical techniques in machine learning, such as multivariate analysis and time series analysis, enhance model performance and data insight.

- Descriptive statistics are crucial for initial data analysis. They help in model selection and data preprocessing by identifying key data characteristics.
- Understanding and applying both descriptive and inferential statistics is fundamental in machine learning, as it allows us to build robust models and make informed decisions.

12

SPECIAL TOPICS IN MATHEMATICAL AI

Game Theory and AI

Game theory is a fascinating and rich field of mathematics that plays a crucial role in artificial intelligence, particularly in strategic decision-making scenarios. At its core, game theory involves the study of mathematical models of conflict and cooperation between intelligent, rational decision-makers. It's not just about games in the traditional sense; it extends to economics, politics, and beyond—anywhere strategic interactions occur.

In the context of AI, game theory provides a framework for developing algorithms that can make decisions in competitive environments. For instance, consider two AI systems playing chess. Each system must evaluate potential moves based on immediate benefits and how these moves might allow the opponent to respond. This strategic depth is modeled perfectly within the frameworks of game theory.

One of the fundamental concepts in game theory is the Nash Equilibrium, named after mathematician John Nash. This concept benefits AI as it represents a state where no player can benefit by changing strategies. In contrast, the other players' strategies remain unchanged. Reaching a Nash Equilibrium in AI can mean finding an optimal strategy given the opponent's strategy.

Another critical area where game theory intersects with AI is in the design of multi-agent systems, where multiple AI agents interact and make decisions independently. These systems are prevalent in scenarios ranging from automated trading systems in financial markets to cooperative tasks like synchronized drone flying.

Moreover, game theory also aids in the ethical training of AI systems. By simulating various decision-making scenarios, developers can teach AI the consequences of different actions, guiding them to make choices that are both optimal and ethical.

In essence, game theory equips AI with the ability to predict and strategize, making it indispensable in developing intelligent systems operating in dynamic and competitive environments. As AI continues to evolve, integrating game theory within its algorithms remains a vital area of research, promising more intelligent and intuitive AI systems in the future.

Information Theory

At its core, Information Theory is concerned with quantifying how much information is in a message and devising efficient ways to encode and transmit it.

The foundational concept here is entropy, introduced by Claude Shannon. Entropy measures the uncertainty or the average amount of information a stochastic data source produces. Think of it as a way to gauge the unpredictability of information content. For instance, if you toss a fair coin, the outcome is highly uncertain, leading to higher entropy. In contrast, a biased coin would have lower entropy because the outcome is more predictable.

Another key element in Information Theory is mutual information, which measures the amount of information that one random variable contains about another. This concept is crucial in feature selection in machine learning models, where we aim to reduce redundancy and enhance the predictive power of the models by selecting features that provide the most unique information about the target variable.

Information Theory also extends to channel capacity, the maximum rate at which information can be reliably transmitted over a communication channel. This principle is fundamental in data transmission technologies, from telecommunication networks to data compression algorithms.

In the context of AI, Information Theory helps in understanding and designing learning algorithms that can efficiently process, compress, and interpret data while reducing information loss. It's particularly useful in areas like neural networks, where optimizing information flow can significantly enhance performance.

By integrating these principles, AI developers can create more efficient algorithms capable of handling vast amounts of data, making accurate predictions, and ultimately driving forward the capabilities of artificial intelligence systems. Information Theory not only supports the technical framework of AI but also enhances our understanding of how information is processed and optimized in complex systems.

Chaos Theory and AI

Chaos theory, a fascinating branch of mathematics, explores how even simple systems can exhibit unpredictable behaviors under certain conditions. This theory is particularly relevant to artificial intelligence (AI) as it helps understand how small changes can drastically affect the outcome of complex systems, as is often the case in AI models.

Consider an scenario where a slight alteration in input

data can lead to significantly different outcomes. This is reminiscent of the butterfly effect, a popular concept in chaos theory that suggests that the mere flap of a butterfly's wings could ultimately cause a tornado halfway across the world. Similarly, minor tweaks in AI's data or algorithm parameters can sometimes result in unexpectedly large differences in the model's behavior or predictions.

This sensitivity to initial conditions is why AI developers must be meticulous in setting up their models and choosing their data sets. I learned this lesson the hard way during my early days in AI research. I once spent weeks trying to understand why my model's performance was fluctuating wildly, only to discover that minor, seemingly insignificant changes in the data preprocessing were to blame. This experience was a practical demonstration of chaos theory, highlighting how small changes can lead to big impacts.

Incorporating chaos theory into AI involves more than acknowledging this sensitivity; it requires robust design and testing strategies to ensure that AI systems can handle and adapt to the inherent unpredictability of real-world data. Techniques derived from chaos theory can also enhance the interpretability and resilience of AI systems, making them more reliable and trustworthy.

Understanding chaos theory thus provides AI practitioners with valuable insights into the complexity and dynamics of the systems they are working with. It teaches them to anticipate and mitigate potential issues arising from chaotic behaviors, ensuring that AI systems perform consistently and effectively in various conditions. This integration of chaos theory into AI improves model performance and contributes to the field's advancement towards more adaptive and intelligent systems.

Graphical Models

Graphical models are a cornerstone of artificial intelligence, particularly when it comes to understanding and managing probabilistic relationships in complex systems. These models, which include Bayesian networks and Markov random fields, provide a structured representation of the joint probability distributions over a set of variables. This makes them incredibly useful for tasks that involve uncertainty and reasoning under incomplete information.

Let's break it down a bit. Imagine you're trying to predict the weather. You have various factors like temperature, humidity, and wind speed. Each of these elements influences the other in some way. Graphical models help by mapping out these dependencies in a visual format, often using nodes to represent the variables and edges to denote the relationships between them. This visualization isn't just for looks; it allows algorithms to perform inference and learning efficiently.

Bayesian networks, a type of graphical model, are directed acyclic graphs where each edge has a direction, and no loops are formed. They are particularly good at representing causal relationships. For instance, smoking might increase the likelihood of developing lung cancer, which could lead to an increased risk of mortality. Each of these variables would be a node in the network, with directed edges representing the causal influences.

On the other hand, Markov random fields are undirected graphical models. They are excellent for representing variables that mutually influence each other, such as the pixels in an image for computer vision tasks. In these models, every node is conditionally independent of the others, given its neighbors. This property is beneficial in image recognition, where the value of a pixel is highly dependent on the adjacent pixels.

The math behind graphical models is deeply rooted in probability theory and statistics. Understanding concepts like conditional independence, joint probability distributions, and Bayes' theorem is crucial. These models are not just theoretical constructs but have practical applications in various fields, including genetics, where they are used to study the inheritance patterns of genes, and machine learning, where they are used to build algorithms that can learn from and make decisions based on data.

In essence, graphical models provide a framework that melds probability theory with graph theory, offering a powerful tool for explaining and quantifying the uncertainty and interdependencies in different data types. They are indispensable for anyone venturing into AI's toolkit, providing clarity and insight into complex probabilistic models.

Reinforcement Learning

Reinforcement learning is a fascinating area of study that sits at the intersection of machine learning and decision-making. It involves training algorithms to make decisions by rewarding desired behaviors and punishing undesired ones. This method allows machines and software agents to automatically determine the ideal behavior within a specific context to maximize performance. Simple, right? But let's dig a bit deeper.

At the core of reinforcement learning is the concept of the agent. This agent interacts with its environment, which is defined by a state. The agent makes decisions or actions based on the state it's in, aiming to achieve a goal. Each action results in a reward or a penalty, and the state of the environment changes in response to the action.

The mathematical backbone of reinforcement learning involves several key concepts from probability and statistics and algorithms that can calculate the best course of action from each state. One of the fundamental algorithms used in reinforcement learning is the Markov Decision Process (MDP). An MDP provides a framework for modeling decision-making situations where outcomes are partly random and partly controlled by a decision-maker.

MDPs are defined by:

1. The environment could be in A set of states (S).
2. A set of actions (A) the agent can take.
3. A transition function (T) that predicts the next state, given a current state and an action.
4. A reward function (R) that gives immediate rewards or penalties after state transitions.

An MDP aims to discover a policy for the agent: a strategy of choosing actions in given states so that the agent maximizes the total reward it receives over time. This often involves calculations of expected values and dealing with probabilistic outcomes, which is where our good friend, mathematics, comes into play.

Dynamic programming techniques, such as value iteration and policy iteration, are commonly used to solve MDPs. These methods involve iterating over possible policies and improving them based on the expected utility of taking actions in given states.

Another critical concept in reinforcement learning is the exploration-exploitation trade-off. This dilemma forces the agent to decide whether to explore new actions to discover better long-run rewards or exploit known actions that give the best immediate reward. Balancing this trade-off is crucial for effective learning. It is often managed by algorithms like ε-greedy, where the parameter ε helps manage the level of exploration.

Reinforcement learning isn't just theoretical; it has practical applications in robotics, automated trading systems, and gaming. For instance, it has been famously applied to teach computers to play and excel at complex games like Go and chess, surpassing human expert performance.

In summary, reinforcement learning is a powerful method for teaching machines to make decisions based on a solid foundation in mathematical concepts. It combines elements of trial and error with sophisticated mathematical frameworks to solve problems that involve making a sequence of decisions toward a goal.

Quantum Computing and AI

At its core, quantum computing departs from classical computing by using quantum bits, or qubits, which can exist simultaneously in multiple states thanks to the superposition principle. This capability allows quantum computers to process many possibilities simultaneously, making them exceptionally powerful for certain types of computation.

The implications of quantum computing for AI are significant. Traditional computers, which operate using bits that must be either 0 or 1, can be outpaced by quantum computers in tasks involving large, complex datasets and scenarios requiring immense computational power. For instance, machine learning algorithms, which require the processing of large amounts of data and complex pattern recognition, could be dramatically accelerated.

Quantum algorithms, like Shor's algorithm for factoring large numbers and Grover's algorithm for database searching, provide glimpses into potential speedups. These algorithms show that tasks taking classical computers exponentially longer might be done in polynomial time on quantum machines. This speed could revolutionize fields such as cryptography and complex system simulation.

However, integrating quantum computing with AI also presents substantial challenges. The hardware for quantum computers is relatively easy to manage and scale because qubits are extremely sensitive to their environment. Any slight change in temperature, electromagnetic fields, or even cosmic rays can cause a qubit to lose its quantum properties in a phenomenon known as decoherence.

Moreover, the mathematical foundations of quantum computing are deeply rooted in linear algebra, particularly in manipulating vectors and matrices in complex vector spaces. Understanding these principles is crucial for developing and programming quantum algorithms. For AI practitioners, a solid grounding in these areas of mathematics is essential to harness the potential of quantum computing.

Despite these challenges, the ongoing research and development in quantum computing suggest a promising future where AI can leverage this technology to achieve what's currently considered impossible. As we continue exploring this frontier, the synergy between quantum physics and AI is the key to the next great leap in our technological capabilities.

Chapter Summary

- Game theory is essential in AI for strategic decision-making, extending beyond traditional games to economics and politics.
- It helps AI algorithms make decisions in competitive environments, such as chess, by evaluating moves strategically.
- Nash Equilibrium in game theory is crucial for AI to find optimal strategies in competitive settings.
- Game theory is also used in multi-agent systems, ethical AI training, and teaching systems to make optimal and ethical decisions.
- Information Theory focuses on quantifying message information and improving data transmission efficiency.

- It introduces concepts like entropy and mutual information, which are vital for feature selection in machine learning and for optimizing data transmission.
- Chaos theory in AI highlights how small changes in input can lead to significant differences in outcomes, emphasizing the need for meticulous model setup.
- Graphical models in AI manage probabilistic relationships and are used in various applications like weather prediction and genetics.

THE FUTURE OF MATH IN AI

Current Trends and Future Directions

One of the most compelling trends is the growing emphasis on topology and differential geometry in machine learning algorithms. These mathematical fields offer profound insights into data's complex geometrical and topological structures, which are crucial for developing deep learning models that can handle increasingly intricate tasks.

Moreover, data availability and a surge in computational power have led to a renaissance in Bayesian methods. These techniques, rooted deeply in probability and statistics, are vital for making decisions under uncertainty and are being employed more extensively in AI to improve decision-making processes.

Looking forward, quantum computing presents an

intriguing frontier for AI. The principles of quantum mechanics are expected to revolutionize our computational capabilities, leading to new algorithms that could solve problems beyond our reach. The mathematical challenges in this domain are immense, but so are the opportunities promising to unlock new potentials in AI capabilities.

In essence, the future of math in AI is not just about advancing what we already know but venturing into uncharted territories, where the synergy between mathematical rigor and AI innovation can lead to transformative breakthroughs. As we continue exploring these frontiers, the importance of a solid mathematical foundation in AI becomes more apparent, underscoring the themes explored throughout this book.

Ethical Considerations in AI

As we dive into the ethical considerations of AI, it's crucial to recognize how deeply intertwined mathematics and ethics are in this field. The algorithms that power AI systems are built on mathematical models, which inherently carry the biases and values of their creators. This raises significant ethical questions, especially regarding fairness, privacy, and accountability.

Firstly, fairness in AI is a significant concern. Mathematical models can perpetuate or even exacerbate societal biases if not carefully scrutinized. For instance, a recruitment AI that uses historical hiring data may learn and perpetuate biases against certain demographic groups. Fairness requires rigorous mathematical analysis to identify and mitigate these biases, often involving complex statistical methods and ethical decision-making frameworks.

Privacy is another critical ethical issue. AI systems frequently rely on vast amounts of data, including sensitive personal information. The mathematical techniques used to process and analyze this data, such as encryption and differential privacy, play a pivotal role in safeguarding individual privacy. However, these methods must be refined to keep pace with evolving technologies and privacy concerns.

Accountability in AI refers to the ability to trace and justify decisions made by AI systems. This is inherently linked to the transparency of the mathematical models used. Often, the complexity of these models can make them "black boxes," where it's difficult to understand how inputs are transformed into outputs. Developing mathematical techniques that promote transparency and interpretability is essential for building trust and accountability in AI applications.

Moreover, the global impact of AI technologies necessitates a diverse approach to ethical considerations. Different cultures and societies may have varying expectations and norms, which should be reflected in the mathematical models to ensure culturally sensitive applications.

In conclusion, the future of math in AI is not just about advancing technology but also about enhancing our ethical frameworks. As mathematicians and AI developers, we are responsible for ensuring that our AI advancements are not only technically sound but also ethically robust. This will require ongoing dialogue, interdisciplinary collaboration, and a commitment to understanding the broader implications of our work.

Interdisciplinary Approaches

As we look toward the future of mathematics in artificial intelligence, one of the most exciting developments is the increasingly interdisciplinary approach to AI research and application. This convergence of different fields enhances AI's capabilities and broadens the scope of problems it can tackle.

Take, for instance, the integration of biology and mathematics through computational biology, revolutionizing our understanding of complex biological systems and how diseases operate at a cellular level. AI models incorporating mathematical biology are now being used to predict how cancer cells evolve and respond to treatment, potentially leading to more personalized medicine strategies.

Similarly, the fusion of psychology with AI,

particularly in cognitive computational models, deepens our understanding of human behavior. By applying mathematical principles to model cognitive processes, AI can simulate human-like decision-making, offering insights into the underlying mechanisms of the brain. This interdisciplinary approach advances AI and provides a valuable tool for psychological research.

In environmental science, AI and mathematical modeling are being used to tackle climate change. These models help predict weather patterns, assess the impact of natural disasters, and manage resources more efficiently. The ability to analyze vast datasets through AI algorithms enhances the accuracy of these models, making them invaluable in our fight against environmental challenges.

Moreover, the arts have been included. AI makes significant inroads into creative processes through algorithms that learn and mimic artistic styles. By understanding the mathematical foundations behind art, such as patterns, symmetry, and geometry, AI can assist in creating complex artworks, music, and literature, opening up new avenues for creative expression.

This interdisciplinary approach has challenges, primarily the need for experts in one field to acquire a working knowledge of other areas. However, the benefits, including innovative solutions and a more comprehensive understanding of complex problems, make these challenges worth tackling.

As AI continues to evolve, the integration of diverse

disciplines through the lens of mathematics promises to enhance AI's capabilities and transform how we approach and solve the myriad challenges facing our world today. This collaborative future is not just a possibility—it is already unfolding, and its potential is boundless.

Continued Learning Resources

As we wrap up our exploration of the fundamental mathematics necessary for AI, it's crucial to remember that learning is a continuous journey. The landscape of artificial intelligence is ever-evolving, and staying updated with the latest mathematical tools and theories is essential for anyone looking to remain relevant in this field.

One of the best resources I've found invaluable in my journey is the plethora of online courses available. Platforms like Coursera, edX, and Khan Academy offer courses in everything from introductory algebra to more advanced topics like machine learning and statistical theory. These platforms often include courses created by leading universities and companies, ensuring high-quality, up-to-date content.

Books, too, are an indispensable resource. Whether it's a classic text like "Introduction to Linear Algebra" by Gilbert Strang or more specialized works like "Pattern Recognition and Machine Learning" by Christopher Bishop, the depth of knowledge in these books is vast. I still remember the long nights spent with Strang's book during college, trying to wrap my head around complex vector spaces. It was challenging, but those moments of finally understanding a concept were profoundly satisfying.

Additionally, academic journals and conferences are gold mines for the latest research and developments in the mathematical foundations of AI. Journals like the Journal of Machine Learning Research and conferences like NeurIPS (Conference on Neural Information Processing Systems) provide insights into cutting-edge research and the future direction of AI technologies.

For a more interactive learning experience, discussion forums and online communities such as Stack Exchange, Reddit's r/math, and r/MachineLearning can be beneficial. These platforms allow you to pose questions, exchange ideas, and receive feedback from peers and experts worldwide.

Lastly, I always appreciate the power of networking. Attending seminars, workshops, and meetups can provide opportunities to learn from experienced professionals and make connections that could be invaluable in your career.

Remember, the journey of learning never truly ends. Each step you take builds upon the last, opening new doors and presenting new opportunities. Whether you're a student just starting or a professional looking to refine your skills, the resources are out there. It's up to you to take the next step.

Final Thoughts

As we close the pages of this exploration into the basic math essential for AI, it's important to reflect on our journey and the road ahead. The landscape of artificial intelligence is ever-evolving, and the role of mathematics in this field cannot be overstated. Mathematics is the backbone of innovation, from algorithms that learn to recognize patterns to systems that make decisions with minimal human intervention.

The future of math in AI promises even greater integration and sophistication. As computational power increases and we delve deeper into uncharted territories like quantum computing and advanced machine learning models, the demand for a solid mathematical foundation intensifies. The challenges ahead are not just technical but also ethical and philosophical as we harness these powerful tools to shape our world.

Anyone entering this field will need to embrace continuous learning and stay abreast of new developments. The journey through the mathematical landscapes we've explored—algebra, calculus, statistics, and beyond—provides the toolkit necessary to navigate the complexities of AI. Yet, the learning should not stop here. As AI continues to permeate various aspects of life, the intersection of math and AI will likely spawn new areas of study and specialized applications, making lifelong learning beneficial and essential.

In conclusion, the fusion of math and AI holds immense potential to drive progress across industries and societies. Armed with the knowledge from this book and an unyielding curiosity, you are well-equipped to be part of this exciting journey. The future is a canvas for your mathematical creativity and the innovations it will bring to artificial intelligence. Let's step forward with the resolve to use this knowledge responsibly and innovatively for the betterment of humanity.

ABOUT THE AUTHOR

Andrew Hinton is a prolific author specializing in Artificial Intelligence (AI). With a background in computer science and a passion for making complex concepts accessible, Andrew has dedicated his career to educating others about the rapidly evolving world of AI. His debut series, AI Fundamentals, is a comprehensive guide for those seeking to understand and apply AI in various professional settings. Andrew's work caters to a broad audience, from managers to coders, breaking down AI basics, essential math, machine learning, and generative AI clearly and engagingly. His ability to demystify the complexities of AI has made him a trusted voice in the tech industry. Andrew's work imparts knowledge and empowers his readers to navigate and innovate in an AI-driven world.

READER BONUS

Thank you for purchasing this book! Get your copy of Andrew Hinton's limited edition book titled "The Power of AI."

It's a must-read for anyone interested in the future of technology and its impact on our world.

Visit bookboundstudios.wixsite.com/andrew-hinton or scan the QR code below.

SCAN ME

Made in United States
North Haven, CT
10 November 2026

22106123R00104